DOES COFFEE CAUSE CANCER?

DOES COFFEE CAUSE CANCER?

AND 8 MORE MYTHS ABOUT THE FOOD WE EAT

DR. CHRISTOPHER LABOS

Published by ECW Press
665 Gerrard Street East
Toronto, Ontario, Canada M4M 1Y2
416-694-3348 / info@ecwpress.com

Cover design: Jessica Albert

LIBRARY AND ARCHIVES CANADA CATALOGUING IN PUBLICATION

Title: Does coffee cause cancer? : and 8 more myths about the food we eat / Dr. Christopher Labos.

Names: Labos, Christopher, author.

Description: Includes bibliographical references.

Identifiers: Canadiana (print) 20230460437 | Canadiana (ebook) 2023046047X

ISBN 978-1-77041-722-9 (softcover)
ISBN 978-1-77852-200-0 (ePub)
ISBN 978-1-77852-202-4 (Kindle)
ISBN 978-1-77852-201-7 (PDF)

Subjects: LCSH: Nutrition—Popular works.

Classification: LCC RA784 .L33 2023 | DDC 613.2—dc23

This book is funded in part by the Government of Canada. *Ce livre est financé en partie par le gouvernement du Canada.* We also acknowledge the support of the Government of Ontario through the Ontario Book Publishing Tax Credit, and through Ontario Creates.

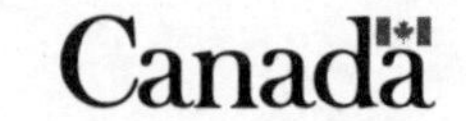

PRINTED AND BOUND IN CANADA

PRINTING: MARQUIS 5 4 3 2

To my family

For their unending support and encouragement
in all aspects of my life

Contents

Preface

This book is a fictional story about real things that aren't true. It's a story about food and being healthy and why that seems so complicated these days. It's a story about why eggs went from being bad for you, to being good for you, and then maybe bad for you again. It's a story about why coffee was declared a carcinogen and why a hot dog was supposed to be as bad as a cigarette. It's a story about all these things because all these things are real. You probably saw these stories on the news or read about them online. But much of what you hear or read about food research is contradictory and wrong. When you read this book, you'll understand why.

The story you'll read in this book didn't happen. The main character isn't me. I never met somebody at the airport on my way to a medical conference who asked about vitamin C and set off a chain of events that changed my life. My work travel isn't nearly that exciting. But the stories are kind of true in that, at some point, in some way, these types of conversations do happen. Somebody will tell you that red wine is good for your heart or that dark chocolate has health benefits that can boost your brain power. If I'm within earshot, I usually tell them why it's not true. I don't get invited to many dinner parties for that reason.

I like to debunk things. That's what I do. I see things that don't make sense, and I point out to people why they don't make sense. Kind of like Socrates, only with a hopefully better long-term outcome. This is what epidemiology is. That's what epidemiologists do. During the pandemic, I had a hard time explaining to people that epidemiology was not the same thing as infectious disease research. Epidemiology is the study of disease, but at the population level, and it can encompass anything from heart disease to cancer to arthritis. The critical part is to understand how to design a study properly. If you do it well, you will get an answer to your question. Do it wrong, you start thinking that coffee causes cancer.

This was my goal: I wanted to write a book about epidemiology and study design and medical statistics. I then quickly realized how boring that sounded, so I decided to write a book about something everyone cares about. I wrote about food. Everyone eats food and everyone wants to eat healthy (most of the time), but most people are confused about how to go about it. It seems like everything causes cancer. But then you hear that the same food prevents cancer. It would be understandable if you were confused.

One study reviewed how different foods have been evaluated in medical research over the years. The findings were a bit all over the place. If you, for some reason, wanted to sit down and read every study ever done about whether potatoes increased your risk of cancer, you might not be too certain where things stood by the time you were done.

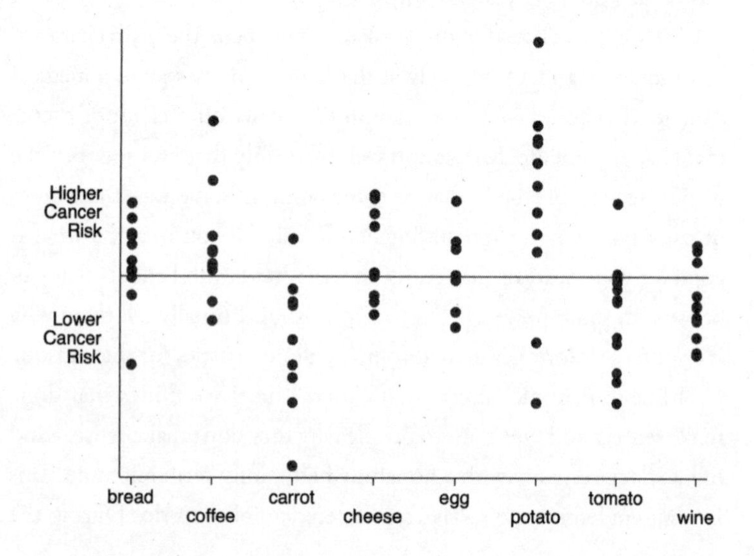

Medical studies are like movies. Not all of them are good. Some are interesting, some are definitive, and some are just bad. It's okay for studies to be wrong sometimes, but it's a problem when we cling to old ideas that should have been jettisoned long ago. The French paradox and the notion that red wine is good for your heart was a provocative and interesting idea in the 1980s. It's not so relevant 40 years later.

This isn't a diet book. I'm not going to tell you what to eat. You can eat whatever you want. A long time ago I made a promise never to write a diet book. When I first started doing science communication, I did a radio interview with Tommy Schnurmacher at CJAD radio in Montreal. He told me I could make a lot of money writing a diet book because a year later I could write a different diet book that completely contradicted the first one and double my money. He was, I think, joking, but he is under instructions to track me down if I ever do write a diet book, so that's not what this is.

This is a story with a beginning, middle and end, but within it, there's a lesson about food. It's a lesson about how research can sometimes go wrong, why we believe things that aren't true, and how we can maybe do better if we try. It's a story about epidemiology and medical research and how abstract things apply to the food you eat every day.

Oh, yeah, it's also a love story.

MYTH #1

Vitamin C Fights the Common Cold

I don't like to fly. But there's no practical way to travel to California other than by plane, and I didn't quite have the luxury of the time it takes for a multi-day road trip across the continent. So I found myself in an airport lounge trying to figure out the ideal time to take Gravol.

When I was young, I got airsick on planes, and there's still something about the smell of airplanes that bothers me. I also don't like the food, being packed into my seat like a sardine, or the minimalist bathroom that is often a few shades off my ideal for cleanliness after the first hour of the flight. All this is to say, I don't relish air travel. The trick, I eventually learned, was to medicate myself about 30 minutes before takeoff. Not only does it quell my airsickness, it also knocks me out and lets me sleep through most of the flight.

So there I was in the airport lounge, doing some quick math in my head, when the man next to me tapped me on the shoulder and asked me a question. Years later, I wondered what would have happened if he had gone up to someone else, how that weekend might have played out differently as a result. But by fate or luck, he came up to me.

"Is that cold medicine, by any chance?" he asked.

He didn't inspire confidence. His hair was messy, and he was flushed with a light tinge of sweat, possibly from a mild fever. His nose had a reddish glow, which suggested either chronic alcoholism or a recent cold. I gave him the benefit of the doubt and assumed the latter. When he sneezed, I had my answer, although the croakiness of his throat was a good clue too.

"No," I said. "It's for airsickness."

He asked if I had cold medicine. I unzipped the front pouch of my carry-on and fished out the small bottle of acetaminophen, opened the top and shook two tablets into his waiting outstretched palm. He thanked me and popped the tablets in his mouth and swallowed them without water, not even looking at the pills. I could have poisoned him and committed the perfect crime. I shook off the thought. I'd been reading too much Agatha Christie and P.D. James.

"You're a lifesaver." He held out his hand and told me his name was Jim.

I looked at his outstretched hand, the hand he'd just sneezed into. I didn't want to be rude, but I didn't want to shake it. "You shouldn't take pills from strangers."

Jim's face froze.

"I mean that was just generic acetaminophen, but . . ."

"Acetamino . . . ?" he stumbled.

I clarified that acetaminophen is the chemical name for Tylenol, a brand name. He asked if I was a chemist. When I told him I was a doctor, he seemed skeptical. Apparently, I didn't look like a doctor. Maybe he thought I looked too young. I might be pushing 40, but I figured I could pass for a few years less. Denial is a powerful thing.

It took some doing, but I convinced him, and he relaxed.

He obviously assumed that the Hippocratic oath offered some protection against random airport poisonings, which to some degree it probably did. "I need some vitamin C," he said. "I don't know if I'm really sick, but I figure I should treat it and be safe."

The voice in my head said, "Don't engage! Don't engage! This is a natural break point for the conversation. Look down at your phone and wait out the clock until the plane starts boarding." It was good advice. But I don't always listen to that voice, even if it does sound like Michael Caine and carries all the implied authority of a British accent. It's hard for me to let stuff like that go.

"You know, vitamin C doesn't really help fight off the common cold."

"Yes, it does. People use it all the time."

"I didn't say people don't use it. I said it doesn't work."

He frowned. "How could companies sell it if it didn't work?"

"It's complicated." I didn't want to get drawn into a long conversation. We'd be boarding soon.

"Haven't people tested vitamin C as a cold treatment?"

"They have. I think there have been over 30 studies with maybe over ten thousand people in them."

"So what's the problem? Either vitamin C works or it doesn't."

More data isn't always better, I explained. Sometimes data makes things muddled. He didn't say anything. I hadn't made my point clear. The display above the boarding gate said we wouldn't board for at least 40 minutes. Maybe longer. Outside the gate, our plane hadn't arrived. We might be there for a while. I figured I'd try again. "You've probably heard that if you take vitamin C when you get a cold, the cold goes away faster."

"Of course, that's why I take it."

"Here's the thing though. Taking vitamin C when you get sick does not treat a cold."

"Nuh-uh, does too."

Was he trying to be a jerk?

"There was a study on this . . . I think," he said.

"Actually, there've been seven studies on this question, specifically, on whether vitamin C makes a cold go away faster."

"Even better!"

"They were negative."

Jim frowned. Nobody likes having their worldview shaken, even on relatively trivial issues like the medical benefits of vitamins. He demanded to know where the idea came from. I explained that there are two ways of taking vitamin C.

His eyes widened. "You mean like take it as a pill or as a suppository?"

How was this conversation going to play out? "No, I mean you can take it when you start getting sick or on a daily basis."

He said he took it when he started getting sick. But, I explained, the seven studies showed it wouldn't help him get better faster. He had to take it regularly.

"I mean it's one thing to take it when you get sick, but every day of your life just to prevent a cold?" He said.

"It doesn't prevent colds."

"But you just said . . ."

He spoke so loud people looked our way. Maybe he was thinking about all the money he'd spent over the years taking vitamin C tablets for no apparent benefit. After many long years of dealing with difficult patients, I've gotten used to de-escalation. I explained that groups who took vitamin C and those who didn't had the same number of colds. According to medical research, the only benefit to taking vitamin C regularly was that when you got a cold, the symptoms went away a bit faster.

How much faster? he asked. By a few hours give or take, I replied. He seemed less than impressed, and I couldn't disagree. Cochrane — an international group of researchers that reads and summarizes medical data on specific subjects — looked at 31 studies and reported that vitamin C reduced the duration of cold symptoms in adults by about 8 percent and in children by 14 percent.

When Jim heard this, he tried to do the math. From all outward appearances, it was not going well, so I spelled it out. If you got a cold that lasts for four days — let's say 100 hours for simplicity — and you were taking vitamin C regularly when you got sick, then you'd reduce the duration of your cold from 100 to 92 hours, or 8 hours shorter.

Jim still didn't think it was worth it, and I couldn't argue with him, so I just nodded.

"But people do say that vitamin C prevents colds," he said.

"There was one group of people where that did happen."

"Ah good. Who?"

He seemed pleased. Completely abandoning something you believe in is hard for people to do. "Marathon runners, skiers, and Canadian soldiers performing subarctic military maneuvers."

"Marathon runners, skiers, and . . . what?"

I repeated what I'd just said.

"Canada has its own army? Oh, right, the Mounties."

"No, that's the RCMP, which is the national police force that . . . It doesn't matter. The point is —"

"So there are situations where vitamin C works."

He'd regained some of his enthusiasm for the medical benefits of what I assumed was his favorite vitamin. "Are you a marathon runner, skier, or a Canadian soldier?" I asked.

He looked down at the slight belly peeking over his belt. He wasn't in bad shape, but he probably wasn't an elite athlete either. "No, I'm not good with the cold," he said. I doubted if that was the real limiting factor. "What does that mean for me as a non-runner-skier-soldier?" he asked.

"That's tricky. People who are very enthusiastic about vitamin C —"

"Why would people be enthusiastic about vitamin C?"

"Anyone selling you vitamin C is going to be bullish about its health benefits because of . . . let's just say enlightened self-interest." Greed was another word for it. But vitamin C's advocates claim it helps people who are exposed to periods of brief physical stress. Jim brightened at this. A drowning man will clutch at anything.

"That makes sense. When your body's stressed, your immune system is weakened, and vitamin C could give you the boost you need to fight off an infection."

He was pleased with that reasoning, but I burst his bubble when I said it could just be due to random chance. Most people don't love talking about statistics as much as I do, and why would they, but random chance is a serious problem in medical research. I decided to explain the concept by telling Jim about ISIS-2.

"The terrorist group?"

"What? No." This was not an auspicious beginning. I told Jim the ISIS trials were a series of four cardiology trials that ran from the early '80s to the early '90s. They tested different medications and treatments in patients who had just had a heart attack.

"Not the best name," he said.

He wasn't wrong. But the trials predated the terrorist group by a few decades. "It's actually an acronym for International Studies of Infarct Survival."

"Infarct meaning heart attack?"

"Yes! Exactly." I was impressed. "The second of these studies, ISIS-2 —"

"Maybe don't say that name so loud. Things can get out of hand quickly in an airport. They tase people first and ask questions later."

I lowered my voice and explained that ISIS-2 was the study that showed a benefit to taking aspirin after a heart attack. Jim, like most people, assumed aspirin has been the cornerstone of cardiology treatment for much longer. But the interesting thing about the study, I told him, was that it revealed a subgroup of people immune to aspirin's effects. Geminis and Libras.

"What!" Jim frowned. I think he was starting to worry that I might be a crazy person. Which from his perspective was perfectly possible given I was a random person he'd started talking to in the airport.

"Zodiac signs?" he asked. "Should I start reading my horoscope to get health advice?"

"I don't think that's going to be helpful." I explained that the researchers did it on purpose to make a point about subgroup analyses.

"Kind of a diva way to make a point."

Again, he wasn't wrong. The researchers wanted to demonstrate how studies can go astray. Studies answer one main question. Does a new medication work? Or is it better than an existing medication? But afterward, many researchers, if not most of them, do what are called subgroup analyses. They look at a subsection of the data and ask questions: Does the medication work only in men? Does it work in people with diabetes, or in people with renal failure?

Jim didn't see the problem with that. And there isn't one. It's important to not assume that medication works equally well in all people. But things can be taken too far, I told him. When you design a research study, you set out certain parameters and decide in advance what you're going to look for and how much of a margin of error you're willing to tolerate. Ideally, it would be zero percent, but in reality, it never can be. Researchers have to acknowledge some degree of potential for error. And there are two types of error. If you're studying whether Drug A is better than Drug B, then your type I error is the probability that you'll say that one drug is better than the other when they are the same.

Jim interrupted to ask if he needed to know what the type II error was. When I said no, he was visibly relieved. But even with this reprieve, he didn't seem satisfied.

"I don't understand what you mean by 'random.'" He had no shame using air quotes. "I understand that random, unpredictable things happen. Maybe a machine breaks, or some lab assistant writes down the wrong number in a book, or somebody leaves a window open and a bird flies in and knocks something off a lab bench . . ."

I'm always surprised when I find out how people think science works.

". . . and unpredictable crazy stuff happens all the time. But numbers are numbers. Either something is true or it isn't. How can the numbers be right and random at the same time?"

I thought a visual example might help. "Do you have a quarter?"

"No, I have a cell phone. Why would I carry change in my pocket?"

Fair point. I didn't have any change either. We were going to have to pretend.

"Just imagine you took a quarter from your pocket and flipped it ten times, you would logically get five heads and five tails, right?"

Jim agreed.

"But it's perfectly plausible that you could get six heads and four tails, or maybe even seven and three, or maybe even eight heads and two tails."

Jim didn't say anything.

I took that as tacit agreement. "Now, with statistics, you can figure out that if you flip a coin 1,000 times you should get somewhere between 470 and 530 heads just by random chance. Anything outside that range and it's possible that you have a trick coin."

"How did you do that math so quickly?" asked Jim.

"It's a common example when you teach this to students. But the math isn't important."

"It almost never is."

I bit my tongue. "The point is that you can measure the margin of error you should get from random chance."

"So how do people decide how much error is acceptable?"

That was a debatable point. By convention, the threshold is set at 5 percent. In other words, there's a 5 percent chance you'll say that Drug A is better than Drug B when it isn't.

"That's actually higher than I would have thought," Jim said. "I mean 5 percent is not negligible."

"No, it isn't. But there's a problem here. Remember we talked about the ISIS-2 study and the multiple subgroup analyses where the researchers looked at so many different subgroups and found that Geminis and Libras didn't benefit from taking aspirin after a heart attack?"

"The problem being your horoscope doesn't affect the blood-thinning properties of aspirin?"

"The problem being all this assumes you test just one thing, just the main question you wanted to answer. If there's a 5 percent chance your analysis is wrong if you do one analysis, think about what happens if you run two analyses."

"I wasn't very keen on math in school."

I took my boarding pass and flipped it over to write on the back. "You have to think about it like this. When you do multiple tests, what's the probability that all your tests will be right?"

"How about we save time and you just tell me the answer."

"If you do one test, there's a 95 percent chance the test is right. If you do two tests, then the probability both tests are right is 95 percent times 95 percent."

Number of tests		Chance that all of the tests will be right
1	95%	95%
2	95% x 95%	90%
3	95% x 95% x 95%	86%
5	95% x 95% x 95% x 95% x 95%	77%
10	95% x 95% x 95% x 95% x 95% x 95% x 95% x 95% x 95% x 95%	60%

"And three tests is 95 percent times 95 percent times 95 percent."

"That's the point. The more tests you do, the harder it becomes for all of them to be right." I had started drawing a little table on the back of my boarding pass to make the point clearer.

Jim murmured as he looked over my numbers. "So if you had ten statistical tests, then the likelihood that all ten are accurate is only about 60 percent?" He thought about that for a second. "Which means . . . the chance that you have at least one wrong answer is . . . 40 percent? Is that right?"

I nodded. The chance of a false positive scales up faster than people realize.

Number of tests		Chance that all of the tests will be right	Chance that at least one test will be wrong
1	95%	95%	5%
2	95% x 95%	90%	10%
3	95% x 95% x 95%	86%	14%
5	95% x 95% x 95% x 95% x 95%	77%	23%
10	95% x 95% x 95% x 95% x 95% x 95% x 95% x 95% x 95% x 95%	60%	40%

"Forty percent is a pretty high margin of error," Jim said. "That wouldn't be okay in most businesses."

That was the point the authors were trying to make with the ISIS-2 study and the whole thing about zodiac signs. They wanted to create a ridiculous example that would be impossible to ignore. Astrological signs don't affect how much aspirin thins your blood after a heart attack. They were saying that if you do enough tests, you'll get a bad, meaningless result. It's expected and predictable. I pulled up the original study on my phone and magnified the text to show Jim: "All these subgroup analyses should be taken less as evidence about who benefits than as evidence that such analyses are potentially misleading."

"Shouldn't it be obvious if the results of research are due to random chance?"

"It's really hard to tell, and people have a remarkable blind spot when it comes to random chance. There was this study called the miracle dice study. Its official name was 'The Miracle of DICE Therapy for Acute Stroke: Fact or Fictional Product of Subgroup Analysis?' It's funny and satirical and shows how much havoc the play of chance can wreak on your scientific research."

"I'm skeptical statistics can be funny."

"It had some great one-liners." Great by statistical research standards at least. The study was published in the *British Medical Journal* and the researchers went to a statistics class being given as part of a seminar on strokes. They asked students to help them generate random data by rolling dice and recording the results.

"Couldn't they have done that with a computer?"

"Sure, but I guess they wanted people to see what was happening. Each roll of the dice represented a patient getting a stroke treatment. If you rolled a six, then that meant the patient died. Otherwise, the patient lived."

"Sounds like Dungeons and Dragons. My nephew plays that."

It wasn't a bad analogy because it was kind of like a game. Everybody got white, green, or red dice and had to figure out which dice were loaded. They were told some dice were loaded and some dice were normal. Then they started rolling the dice.

When they analyzed the data, they found the red dice were more likely to produce sixes. But there was a catch. All the dice were normal. Different colors, but perfectly normal. You were no more likely to roll a six than you were to roll a four or a two with any dice. But the participants rolling the dice were convinced the red dice were different.

"The researchers didn't tell them afterward?" Jim asked.

I explained that they had, but the students didn't believe them. Student A, who found that his white die was really good at preventing deaths, was convinced his die was loaded. One of my favorite quotes from the paper is when the researchers talk about what Student A was feeling when he was rolling his die: "When he [Student A] started on the control group he rolled one six, followed by another and then a third. He said that his room felt eerily quiet as he rolled a fourth six: he had never rolled four sixes in a row in his life."

"People don't usually get creeped out by dice," Jim said.

"But he was convinced he had a loaded die. He had forgotten that rolling four sixes in a row is rare, but not impossible."

"And if you have enough people in a room rolling dice then I guess eventually one of them is going to roll four sixes in a row. Kind of like if you have a thousand monkeys working on a thousand typewriters, they will eventually produce the entire works of William Shakespeare."

"Yeah, that's true."

"Unless the monkeys lead an uprising like in *Planet of the Apes*."

"Actually, monkeys are not apes."

Jim said nothing, and I decided not to get sidetracked and pressed ahead. I told him about the authors' warning about the play of chance in medical research: "Chance does not get the credit it deserves. Most doctors admit that chance influences whether they win the Christmas raffle but underestimate the effect of chance on the results of clinical trials they read about. The point is: if you keep analyzing the same data over and over again, looking at this subgroup or that subgroup, you will eventually get fooled into believing something that isn't true."

On hearing this explanation, Jim lit up. "Like aspirin doesn't work if you're a Gemini or Libra or perfectly normal dice were loaded."

"Exactly. If you go looking for a pattern, you'll find it. That was the final joke in the paper, that DICE should stand for: Don't Ignore Chance Effects."

Jim thought *joke* was a liberal interpretation, but he said he understood what I meant. He went off to fill his water bottle from the fountain. When he came back, he said he felt better with the Tylenol I gave him. He said I was a good egg, but he remained skeptical that random chance could blow up medical studies like that.

"Unfortunately, researchers do it all the time. Have you ever heard of Brian Wansink?"

"Is there any reason I should have?" In all fairness, there probably wasn't.

"Have you ever heard about the bottomless soup bowl?"

"Sounds like something a restaurant does as an advertising gimmick."

"Maybe. Wansink constructed this special soup bowl that had a hose that pumped in more soup as people ate. People ate more soup out of it than a regular bowl."

Jim had heard of this. It seemed to jog some vague memory in his brain. It had become a popular piece of research and led to a surge of pop-sci articles suggesting people should buy smaller plates. The idea was if you put less food on your plate, you will eat less. Most of us are conditioned to clear our plates when we sit down to eat. We keep eating even when we're not hungry because we're programmed to eat what's put in front of us. As I told Jim about Wansink's work, he realized that he had heard about many of these research findings even if he was unfamiliar with the name. Wansink and the Cornell Food Lab had produced research showing that kids are more likely to eat fruit if you put fun stickers on them. There was also research where they looked back at every edition of *The Joy of Cooking* and found that serving sizes and calorie counts have been steadily going up over its entire print run. It was popular research, and I quoted it myself.

Jim didn't see the problem, and initially there wasn't one. But things unraveled in 2018 when Wansink wrote a blog post. Jim was impressed he was still bothering to write a blog in 2018. I shrugged. In the blog, Wansink talked about how he told his grad student to keep analyzing a dataset over and over again until they found something interesting. She ultimately did, and they got four scientific papers out of the work. Wansink saw it as an example of hard work and perseverance. But it wasn't. It was p-hacking.

"And what's p-hacking?" Jim asked.

"The p-value is a statistical term that means . . . it's a bit difficult to explain. If the thing you were trying to prove isn't actually true, then how likely is it that you would have gotten the results you saw in your experiment."

"You lost me."

This was going to be complicated. Even many researchers don't fully understand what a p-value is. The actual formal definition is very complicated and somewhat counterintuitive. I tried a couple of explanations before Jim shut down my attempts and asked me if the p-value was just a statistical test that researchers use. I told him it was. He told me he didn't need the details. And in reality, he didn't. The key is understanding that if you keep doing tests over and over until you get a result you like, that's called p-hacking. It also goes by a few different names: data-dredging, multiple hypothesis testing, or salami slicing because you keep cutting up your dataset into thinner and thinner slices of the larger population. You keep analyzing, subdividing, and cutting up your data until you get a p-value that looks interesting and can justify a scientific publication. It's a successful strategy as long as you're willing to ignore all the times when you didn't get the answers you wanted.

I showed Jim a cartoon I use to explain the concept to students.

Wansink, I explained, thought he was giving good advice about not giving up and working hard. But he was actually telling his graduate

CARTOONIST: LUC SPERL

students things that were, to put it diplomatically, not quite up to the highest standards.

"You mean cheating?"

"That's a little harsh. He claimed he didn't realize what he was doing was wrong. But people started looking harder at his research. He had over a dozen studies retracted, and he retired from his university."

"I call BS. I find it very hard to believe that a big name in research, and it sounds like he was a big name, wouldn't know about this."

"Maybe some don't know, don't care, or are willing to cut corners to succeed. But a lot more simply, many researchers don't appreciate the importance of random chance. They do things when they analyze data, and it never occurs to them that some of their findings are just random statistical noise."

"Like with the dice," Jim pointed out. "The students didn't believe the dice were regular dice even after the researchers told them."

"True. Roll enough dice and eventually you'll roll four sixes in a row. If it happened to you, you'd probably think you had loaded dice. But the chance that you would roll the same digit four times in a row is about 0.5 percent. That's obviously not common, but it's not super rare either. It will happen eventually, no matter how shocking it might seem."

"I guess you're right," Jim said.

"It's the same with research. Look at enough subgroups in your data and eventually you'll find one of them has a positive statistical finding in it. And if you aren't aware of the problem, you might think Geminis and Libras shouldn't take aspirin after a heart attack. And if you look hard enough at data that is overall negative, you might convince yourself that marathon runners, skiers, and soldiers benefit from vitamin C."

Someone from the airline announced the plane had arrived and would be boarding soon. People queued up. The gate agent asked them to let priority passengers and anyone needing special assistance board first. The pleas fell on deaf ears because everyone, including Jim and I, knew that the plane had limited space for carry-ons.

"So you think the supposed benefit from vitamin C in this subgroup of people is just random statistical noise?"

"I do. Remember, overall, vitamin C does nothing. It doesn't prevent a cold if you take it regularly, and it doesn't treat a cold if you take it when you get sick. The only protective benefit is in that specific subgroup of marathon runners, skiers, and soldiers."

It was also a fairly small subgroup made up of around 600 people in five studies: three in the ultramarathoners, one in skiers, and one in Canadian soldiers doing subarctic training exercises. Of the five, the skier study was by far the largest and contributed roughly half the people.

"Isn't it weird to combine such different groups together?" Jim asked. "I mean I can understand that marathon runners —"

"Ultramarathon runners."

Jim blinked. "It's hard to see what ultramarathoners have in common with skiers. Competitive skiing is a high-intensity sport, and in the cold —"

"The study in skiers wasn't done on competitive skiers."

"No?"

"It was done in children attending a ski school in the Swiss Alps."

"That's a lot posher than I was picturing in my head."

"It does sound like a really nice vacation."

"If I was given a week-long vacation skiing in the Alps, I wouldn't be under physical stress. I'd be happy, relaxed."

"Thought you weren't big on cold-weather sports."

"I'm more of an après-ski-hot-chocolate-in-the-chalet type of guy. You don't need to ski to enjoy going to Switzerland."

There was some truth to that. We inched toward the desk. Jim mulled things over. "I guess it's plausible that putting your body under extreme stress would suppress your immune system. If you endure extreme physical stress, like athletes or soldiers do over the long term —"

"Except these were short-term studies. The ski school one only lasted a week because that's how long the students were there. The military one ran from the week before to the week after the military exercise and the one in runners was generally about two weeks as well."

Jim wanted to know if there were any long-term studies. There were two. One focused on U.S. marine recruits for two months, the other on competitive swimmers for three months. Both were negative."

"Nobody mentions those in the vitamin C commercials."

"Unsurprising from a marketing perspective."

"So this is just random statistical noise, and I've been wasting my money?" I asked him if he'd heard of the website Spurious Correlations. He hadn't, so I explained that this guy named Tyler Vigen had put together graphs to show how often you can get random, or completely spurious, associations. For instance, Vigen has graphs that show how U.S. spending on science, space, and technology is related to the number of suicides by hanging, strangulation, and suffocation. Or how the number of people who die each year by falling into a pool is correlated with the number of films in which Nicolas Cage has appeared.

I showed Jim the graphs on my phone, and he snickered. "Nicolas Cage!" he said to himself. "I love his movies."

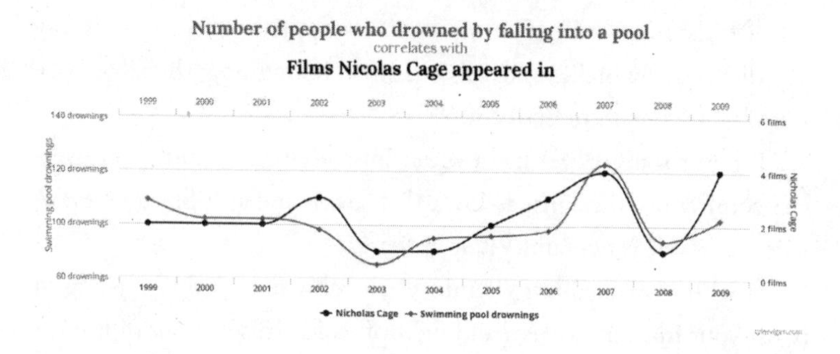

What makes these graphs funny is that they use real data, but the association is random. And if you can find an association between the divorce rate in Maine and per capita margarine consumption, it's not implausible that you could find a link between vitamin C, colds, and soldiers.

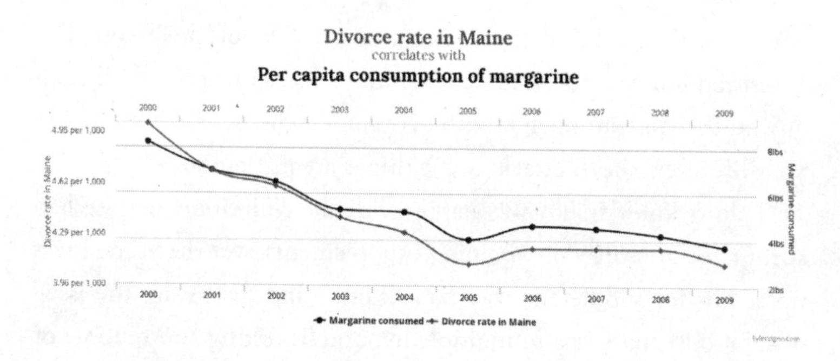

"So why are companies allowed to advertise vitamin C as a cold medication if it doesn't actually work?" Jim asked.

"If you look carefully at the ads, they never outright say that you can take it to fight a cold. They say something like, 'Supports a healthy immune system.'"

"Maybe they do it to compete against the big pharmaceutical companies and their advertising budgets," Jim said. "I don't trust Big Pharma."

"Neither do I. But if vitamin C worked, doctors would prescribe it to patients. Why wouldn't we?"

"Maybe because Big Pharma wants to suppress . . . oh no wait. Big Pharma actually makes and sells vitamins. If anything, they would be pushing doctors to prescribe it."

Jim was right. Pfizer has a whole line of vitamins and other over-the-counter products just as Coca-Cola owns and sells Sprite, Nestea, Minute Maid, Fanta, and Vitamin Water.

"I wish it worked but vitamin C probably doesn't do anything to boost your immune system and fight off cold viruses. A lot of immunocompromised people would benefit if it did. Taking it when you get sick does nothing. The only benefit you see is if you take it regularly and you engage in short-term extreme physical stress."

"Yeah, but I don't run ultramarathons, go to Swiss ski schools, or serve in the Canadian military."

"You could volunteer," I suggested.

"Pass."

"Even if you did, I don't think vitamin C would help you. This is just random statistical noise. Vitamin C doesn't help fight off colds any more than your birth month determines whether aspirin will save your life after a heart attack. Some things are just random."

I don't know if Jim was happy with the realization that he had spent a lot of money on vitamin C supplements over the years. I was upset but for a different reason. I didn't mind discussing the issue of subgroup analyses and multiple hypothesis testing and the role of random chance in medical research. I enjoy this stuff. I was upset because as I walked up the gangway, the smell of the plane made me nauseated. I'd forgotten to take my pill and cursed Jim for coming up to me of all people. But I knew it wasn't his fault. Some things in life are just random.

MYTH #2

Hot Dogs Are as Bad as Cigarettes

The plane was crowded. There were just enough people fighting over just enough overhead space to create a situation that could spill over into violence with little provocation. At my seat, the overhead bins were stuffed with other people's bags. I looked left and right and saw that there was no room to be had. My faith in humanity taken down yet another notch, I looked across the row of seats to the opposite aisle.

Jim waved. He scooched across the row of seats and sat on my right. I wondered if I could pretend that I hadn't seen him. That didn't seem like a successful strategy. Resigned, I put my bag under the seat in front of me.

"What are the odds we would be sitting next to each other?" Jim asked.

"Pretty slim." But from a purely mathematical standpoint, there are only so many seats on a plane. If two people are traveling alone, and if you assume families and groups will want to sit together, and you exclude business class, you're not left with a lot of options. A flight attendant admonished me for not buckling my seat belt. Whatever the probabilities, Jim was sitting next to me and would be for the next few hours. I decided to make the best of it.

"I never asked what you did," I said.

I technically still hadn't, but Jim regaled me with his life as a loss adjuster, which was more interesting than you would think. When somebody is passionate about what they do, it's always interesting when they talk about it. The flight attendant's safety presentation, which always struck me as slightly absurd, interrupted him. Most people know how a seat belt works and if the plane did go down over water, we were done for no matter what. But I keep that type of thinking to myself, especially on planes.

About an hour into the flight, after Jim had taught me more than I ever thought possible about fly fishing, we were served our meals. I peeled back the aluminum foil cover and after some careful reflection decided I was looking at a chicken breast.

Jim looked at this meal, frowned, and murmured something. "I don't know what this is."

I looked over at his tray and understood his confusion. "Couscous?"

"No, that's usually smaller than this."

I pleaded ignorance. When it came to airline food, it was usually best not to set your expectations too high or to ask too many questions. "You ordered the vegetarian option?" I asked. He didn't seem the type.

"No, I think they just gave me the wrong meal." He looked over his shoulder and spotted the flight attendant pushing a trolley slowly down the aisle. He looked like he was going to say something but then calculated his odds of receiving satisfaction and turned back to eye his meal once again.

"It doesn't look too bad," I said.

"Want to switch?"

I didn't. But explaining that diplomatically was going to be tricky.

He looked over his shoulder, but the attendant was even farther away now. "I would have preferred chicken."

A woman to Jim's right said, "Yuck."

"Sorry?" Jim said.

"I don't understand how you can eat meat. I find it disgusting."

Jim didn't seem to know what to say.

"It's the carcass of a dead animal," she said.

She was technically right, so it was a hard point to argue. Claiming that it was tasty would have sounded crass. And eating meat is not the type of thing you can easily justify to someone you just met if you've done it automatically all your life.

"I don't think it's that bad," Jim said.

Jim got no reply. He looked at me for moral support. I shrugged. I try not to get into arguments with people on planes. The industry has a low tolerance for that sort of thing these days. The woman didn't seem to be spoiling for a fight. She introduced herself as Katie and told us she worked for an environmental NGO. She'd seen the impact animal husbandry had on the planet. She also had strong reservations about the ethics of eating animals and showed an intense but genuine concern for their well-being.

"Plus," she added as an epilogue, "eating meat isn't good for you."

"You mean health-wise?" Jim asked.

"Yes!" she said. "It causes cancer. Don't you know that?"

"I don't know . . . I mean . . ."

"Didn't you hear there was a study where eating meat was just as bad as smoking?"

"I . . . I remember something about that. There was something on the news about how eating a hamburger was just as bad as smoking a cigarette."

"I heard it was a hot dog," Katie said.

SOURCE: TWITTER

I remembered those headlines well. They changed depending on what part of the country you lived in. In some parts it was hot dogs, in other places it was hamburgers, and in others, it was smoked meat. But the basic idea was the same. Whatever meat was the local favorite became as bad as tobacco.

I could have kept out of the conversation and read my book. I'd recently started revisiting the Agatha Christie books I'd read when I was young, and I had *The Mirror Crack'd from Side to Side* in my bag. One day I'll learn to keep my own counsel. Jim and Katie went back and forth on this for a bit, but Jim eventually invoked me as a medical expert and wanted to know if hot dogs really were as bad as cigarettes.

"Tell him about the WHO study," Katie said.

"It wasn't a study. It was a report issued by the WHO," I said.

"I'm assuming you mean the World Health Organization and not the band." Jim thought he was being funny.

Katie snorted and rolled her eyes. "You know Bono is vegan."

It was hard to argue with an endorsement from Bono.

"I thought that was an internet myth," Jim said.

"The World Health Organization, not the band." I wanted to get things back on track.

"Well Paul McCartney is vegetarian," Katie said.

It was even harder to argue with a Beatle. I pressed on. "Technically, it was IARC, the International Agency for Research on Cancer, who issued the report. It's one of the WHO's agencies."

Given that most people are not familiar with the inner workings and organizational structure of international agencies, I had to provide some background on the WHO, IARC, and their Monograph program, which was designed to summarize the evidence about possible carcinogens that might pose a hazard to human health. They don't produce new data or conduct experiments themselves. Their function is to review the available evidence and provide a summary for governments, and to some extent, the general population, to make decisions about regulating potentially dangerous and cancer-causing products.

It was going well until I got to the point where I had to explain the difference between hazard and risk.

"I'm pretty sure those are synonyms," Katie said.

"They aren't. Statistically, the concepts of risk and hazard mean slightly different things. The risk of something happening is like asking, 'How likely is it that thing will happen?' Another way to think about it is, if you took one hundred people how many of them would get sick?"

Neither of them said anything, so I kept going. I explained that *hazard* means something a little bit different. It incorporates time into the calculation and looks not just at whether you get sick but also when you would get sick if you did. Jim nodded but looked worried, like someone who was about to lose the thread of the conversation.

"Now the actual mathematical way to explain what hazard means involves calculus and —"

Jim put a hand on my arm. "Hard pass on the calculus."

Katie nodded.

Over the years, I've gotten used to people not wanting me to talk about calculus. It doesn't bother me as much as it used to.

The distinction between hazards and risks is important because sometimes you're more interested in delaying rather than preventing a disease. Cancer is an obvious example. Chemotherapy may not cure your cancer, but treatment might push back a recurrence by six months, or one year or two years. Possibly even longer. Delaying an outcome is still useful even if you can't completely prevent it. High blood pressure is another example. Keep your blood pressure well controlled and maybe you'll still get a heart attack eventually. But having your first heart attack at 70 instead of 50 means you've had an extra 20 years of healthy living.

Unfortunately, IARC uses the term *hazard* to mean something slightly different. They use hazard to mean whether something is dangerous, and they use risk to measure how dangerous it is. We went through all this there on the plane. Jim struggled a bit to understand the whole time-dependent nature of hazard, only to be told that it meant something else in WHO circles.

"I feel like you just wasted five minutes of my life," Jim said.

"To be fair, we are on a plane," Katie said. "You really weren't going anywhere."

"Anyway, that's the difference between hazard and risk," I said.

"I don't understand the distinction," Katie said.

"Think of it this way. A gun is potentially dangerous but not if it doesn't have any bullets in it."

"You probably don't want to say the word *gun* too loudly on a plane," Jim said.

He was right. I lowered my voice. "My point is guns are only dangerous in specific contexts."

"You're big on the Second Amendment, right?" Katie asked.

"Don't get too hung up on the analogy." I thought I might get a better response if I used my go-to analogy of alligators in the zoo. "Alligators are dangerous if you meet them in the wild, but not dangerous if they're behind glass in a zoo."

"I think zoos are unethical," Katie said.

"What about a nature preserve? You know, where they let the animals go free and the humans are behind protective glass or a wire fence or whatever?" Jim asked.

"I guess if the animals are free and unencumbered then that's okay."

I let them go at it. The talk of zoos spilled over into Jim telling Katie about the time his father took him to the zoo, and he saw an elephant for the first time. It was a great story. By the end, I'd resolved to call my father when we landed. The only problem was that Jim forgot why he was telling it.

"Why are we talking about zoos?" he asked.

"I was making a point about hazard and risk," I said.

Jim nodded. "And what was that point again?"

"Context matters and how dangerous something is depends on the situation." How could I explain it better? "Something might be theoretically dangerous without posing a threat to public health on a day-to-day basis. That's what IARC is trying to do. They identify things that are hazardous without making any comment about how dangerous they are."

"I've got a crazy thought," Jim said. "Why doesn't IARC just classify stuff based on how dangerous it is so we can avoid all this wordplay?" Before I could respond, Jim added, "Because wordplay is dumb."

"Won't hazardous things also be risky things?" Katie asked. "Shouldn't they both point in the same direction? And classifying red meat as a Group 1 carcinogen, the same category as cigarettes, implies it's a pretty big deal."

"That's not really what it means," I said.

"No point making these things intuitive and easy to understand," she said.

I ignored the sarcasm. "Group 1 means there's no doubt it's carcinogenic. Group 2A is probably carcinogenic. And Group 2B is possibly carcinogenic. If things are unclear, it gets listed in Group 3 as unclassifiable."

"What about things that don't cause cancer?" she asked.

"That's technically Group 4. Probably not carcinogenic."

She scratched her nose. "What do you mean *technically*?"

"Nothing is actually listed as Group 4."

"They've never found something that wasn't carcinogenic?" Katie asked.

"There were a lot of negatives in that sentence."

She smiled. "You aren't the grammar police. Quit stalling."

"Every product they've reviewed is listed in either Group 1, Group 2A, Group 2B, or Group 3. Nothing is in Group 4."

"How many products have they reviewed and published monographs on?"

"I think a little over a thousand at last count."

Jim snorted. "How do you examine a thousand products, chemicals, ingredients, whatever you want to call them, and not find something that is not cancerous."

It was Jim that got caught on the double negative this time, but I think he got it right.

"To be fair, about half of those are in Group 3, so not classifiable. Basically, they can't say one way or the other."

"But just by dumb luck shouldn't some things come out in Group 4, the group where things don't cause cancer."

Jim made a valid point, but I pointed out to him that this wasn't a random selection of objects. IARC wasn't going around passing judgment on everything under the sun. There was usually a reason they wanted to formally review something like red meat or industrial cleaners. There was some pre-existing doubt or concern.

What really amused Jim and Katie was when I told them there once had been a product listed in Group 4. Caprolactam. It's a chemical used to make nylon and is used in clothing like yoga pants. There was a time when you could confidently state that yoga pants probably didn't cause cancer. However, that time has passed. Caprolactam was re-classified into Group 3 back in 2019.

Katie was dubious about this entire issue. "The World Health Organization isn't sure if me wearing yoga pants is contributing to my risk of getting cancer?"

I wanted to correct her and say "hazard" but that seemed counterproductive and unnecessarily provocative at that point.

"Plus, isn't the fact that I'm exercising a good thing overall?" she continued. "Even if there was a danger to wearing yoga pants, which there obviously isn't because millions of people wear them every day when they go to the gym and are completely fine. Although people wear them as regular pants too at this point."

Jim laughed. "I see people show up to work in sweatpants. I don't know what the world is coming to."

I thought the freedom to show up to work in sweatpants would be amazing, but I held my tongue. The world wasn't ready for such radical ideas.

"I still don't understand how nothing is being classified as not carcinogenic," Jim said.

"It's actually really hard to prove a negative," I said.

"That doesn't make sense," Jim said.

"Prove to me that reindeer can't fly."

"Ummm . . . reindeer aren't real."

"Reindeer are real," Katie said.

I agreed. "Reindeer are real animals. *Flying* reindeer aren't real. Reindeer exist, they just don't fly through the air pulling Santa's sleigh. None of that is real."

A small child shot up from the seat in front of us and looked over the headrest. She had tears in her eyes and her lips quivered. Was this a prelude to a full-blown bawling meltdown? Her mother soothed her and told her not to listen to the crazy people behind them. She shot us a contemptuous look.

"Reindeer are the same as caribou. Caribou is the North American version and reindeer the European version. Neither can fly." I lowered my voice so the girl couldn't hear.

"Glad we could clear that up," Katie said.

"But how would you go about proving that to me if you had to?" I asked.

Jim hesitated. "Common sense?" he suggested.

"It's a useful thought experiment. How would you prove to someone that reindeer can't fly?"

"You could just look at the reindeer, or caribou, and watch them for long enough and you would see that they weren't flying," Jim said.

"Yes, but you can't watch every reindeer, everywhere, at all times. There might be some flying reindeer in some remote location. Like the North Pole."

"Theoretically."

Katie sighed. "This is the dumbest conversation I've ever been a part of. Why did I forget my phone charger?"

"We could conduct an experiment," Jim said. "We could take a reindeer to the top of a very tall building and throw it off and see if it flies then or not."

"So, murder?" Katie asked.

"I didn't . . . I mean . . . I wouldn't actually do it. It's just a theoretical thing."

I felt I needed to jump in and rescue Jim. "The problem with your experiment is you haven't proved reindeer can't fly. You just proved one specific reindeer couldn't fly, or wouldn't fly, at that particular time."

"If you did it a bunch of times with a bunch of different reindeer and none of them ended up flying —"

"And crashed to their deaths," said Katie.

Jim wasn't holding up well under Katie's criticism. But he had touched on an important point about proving a negative. No matter how many reindeer you hurl off a tall building, somebody can always claim that there's flying reindeer you just haven't found yet. That's why you can never prove a negative. "You can't prove that reindeer can't fly. You can only fail to prove that they can. After all, in court, they never find you innocent. They only find you not guilty. Which is why IARC frames things the way they do. It's very difficult to say that something doesn't cause cancer. You can only say there's insufficient evidence to prove that it does."

Katie however was having none of it. "I think this really confuses people because some things obviously don't cause cancer. Chemotherapy doesn't cause cancer. Chemotherapy treats cancer."

"It does, but even things like tamoxifen are included in Group 1 as definitely carcinogenic to humans," I said.

"That's just ludicrous. Doesn't tamoxifen treat cancer? Isn't that the cornerstone of breast cancer treatment and prescribed to millions of women every year?"

Katie was right, and I wasn't going to argue with her because, quite frankly, she intimidated me a little bit. But she made a valid point. Tamoxifen is a very effective treatment for breast cancer. It's an estrogen receptor blocker and since estrogen makes most, but not all, breast cancers grow faster, blocking the estrogen receptor on a tumor cell can be a very effective breast cancer treatment.

It's listed as a carcinogen because while it blocks the estrogen receptor on breast tumors, it stimulates the estrogen receptors in the uterus and increases the risk of uterine cancer. It's a very rare side effect and the cancer risk is only about three-tenths of 1 percent in postmenopausal women. In premenopausal women, the link is more uncertain. For Katie, that was essentially a non-risk. It's not a zero risk of uterine cancer, but it's low enough that everyone agrees the benefits vastly outweigh the risks.

It didn't make sense to Katie that it should be listed as a Group 1 carcinogen. It didn't make sense to me either. But IARC's goal isn't to give treatment recommendations. They aren't trying to determine how dangerous something is. They're trying to judge if a cancer link exists.

"Do they do this a lot?" Katie asked. "List all kinds of nutty stuff as Group 1 carcinogens?"

"Most of the stuff in Group 1 is clearly dangerous. Tobacco, asbestos, X-rays, gamma radiation, stuff like that."

"That seems obvious enough that an international committee of experts didn't need to weigh in."

"But then they'll include estrogen-containing oral contraceptive pills."

"Birth control pills?" Katie asked.

"Yes."

"The World Health Organization's cancer agency . . . listed birth control . . . as a Group 1 carcinogen?"

"Yes."

Katie steepled her hands together. "Don't birth control pills reduce the risk of ovarian cancer? I read that somewhere."

"They do."

"But they're included because . . ."

"The estrogen."

"I see. Because they contain estrogen, and estrogen increases the risk of . . . breast cancer?"

I nodded.

"So because estrogen increases the risk of breast cancer, it gets listed as a carcinogen even though birth control pills prevent the more aggressive ovarian cancer."

"That's about it."

"That's ridiculous."

"It is."

Katie fidgeted in her seat, and I didn't blame her for being baffled by the absurdity of the situation. The risk of breast cancer due to the oral contraceptive pill is vanishingly small. If a woman had a family history or a genetic predisposition, then you might consider an alternative non-hormonal form of contraception. But it's a non-issue for most.

So why bother mentioning it at all? That was Katie's argument. She couldn't understand why IARC, if they were so insistent on drawing a line between the concepts of risk and hazard, didn't just categorize things based on risk. That would have been more meaningful to someone reading their reports. Equating birth control with cigarettes and labeling them both as Group 1 carcinogens is guaranteed to erode your credibility in people's eyes. It did in Katie's at least. In their reports, they go to great lengths to tell people that risk and hazard are not the same thing. They say over and over again people shouldn't compare things within the same group. But that's not what the headlines say when the report gets published.

Katie threw up her hands. "Honestly, I give up. Meat's a Group 1 carcinogen, but I don't know what that means anymore."

"Actually, red meat was listed in Group 2A," I said.

"But earlier you said . . ."

"Processed red meat was listed as Group 1. Red meat itself is in Group 2A."

That distinction was ignored in a lot of the discussion surrounding this issue. IARC made a distinction between processed and unprocessed red meat. Processed red meat meant anything that was cured, salted, smoked, or prepared in some way. It referred to meat products like hot dogs, beef jerky, bacon, salami, sausages, corned beef, and smoked meat. Simply cooking or barbecuing wouldn't qualify something as processed meat. Jim was relieved as he enjoyed barbecuing in the summer. He felt proud of his skills in that area.

"There's an art to grilling a perfect steak," he said.

"No there isn't," Katie said. "You put it on the barbecue and the thing cooks itself. You just have to flip it over once."

If Jim was disappointed, he hid it well. "If the risk is only for processed red meat, then you could also switch to white meat. Something like chicken or veal."

Jim's confusion was understandable. You see many different opinions about what constitutes white meat. But there's a difference between how chefs classify meat and how it gets classified scientifically. In science, red meat comes from mammals. White meat comes from birds. That's where the line is. Chicken is white meat, but beef, pork, rabbit, and venison are all red meat.

Jim remained unconvinced. "I thought meat from a young animal like veal was always classified as white meat because it's healthier?"

"Are we honestly having a conversation about whether it's better to eat baby animals?" Katie asked. "And people wonder why I'm vegetarian."

"What about if it's free-range, grass-fed, or a young animal?" Jim asked.

"None of that matters in terms of the scientific classification. When this stuff gets researched and classified scientifically, it really is as straightforward as mammal vs. bird."

Jim wouldn't back down. "Isn't pork the 'other white meat'?"

"That was a very clever marketing slogan from the 1980s for the National Pork Board. Good marketing but there was nothing scientific about it. It was a commercial."

"Fish?" Jim was desperate.

"Not technically white meat. It gets its own category in food research."

"God, this stuff is complicated."

It was. It's made more complicated because how scientists talk about food isn't how most people talk about food. And while the public interpreted the IARC report to mean meat was carcinogenic, the details of the report made clear, if often ignored, distinctions between red, white, and processed meat. Smoked, cured, salted, or fermented meat, like bacon or hot dogs, had the strongest cancer link. The consensus is that processed meats contain more nitrosamines and polycyclic hydrocarbons. I lost Jim and Katie when I got too deep into the chemistry.

"English please," Jim said.

"Technically that is English," Katie said.

"Maybe just use easier-to-understand terms then."

I took the hint. I think I had made my point. Part of the issue is not so much the meat itself but how it's prepared.

"But what about all the other stuff that's unhealthy about red meat?" Katie asked. "What about the antibiotics they give to animals to make them grow faster?"

That was another complicated issue, but at least we've taken steps to rectify it. "We use a lot of antibiotics in animals. Almost as much as we use in humans. And the more antibiotics you use, the more you contribute to antibiotic resistance."

Katie nodded along in agreement and periodically nudged Jim when I made a particularly damning point. "If people ate less meat, this wouldn't happen as much."

Fortunately, things are better now. You can no longer just walk into any feed store and buy antibiotics for your animals. You need a vet's prescriptions, and that has gone a long way to curtailing antibiotic overuse and abuse.

"And what about growth hormones?" she asked.

"Seems to be on the way out. But it depends on the country. In Canada, it's banned in all animals except non-milk cows."

"Weirdly specific. Why ban it in chickens but not cows?"

"I don't pretend to know how politicians think."

"That's probably for the best," Katie said.

Jim looked back and forth between us to make sure we were done. "So, we're okay then. Meat's not that bad in the end?"

If Jim thought the conversation was over, he was sadly mistaken.

Katie had more to say. "There's still a lot about eating meat that I'm not okay with. We haven't even talked about the environmental stuff yet. Do you know the environmental impact that raising livestock has on the planet? Do you know how many acres of land it takes to raise animals or how many gallons of water?"

I didn't of course, but I was pretty sure that Katie did. I saw Jim close his eyes as he anticipated the onslaught.

"You two realize veal is a baby cow, right? Somebody killed a baby cow so you could eat it." It sounded bad when she put it like that.

I conceded the point. "There's a lot of reasons someone might not eat meat. Environmental, economic, moral, religious, all kinds of reasons. I'm just here to summarize the medical evidence."

"Isn't that what IARC is for?" Jim elbowed me in the ribs.

"I'm more easily accessible."

"We could just read their report."

"Their report is 517 pages long."

"Geez, what do they think they're writing? Harry Potter books?"

"*Deathly Hallows* was over 600 pages," Katie said.

Jim did the math. "Still. I'll just take the executive summary."

It's not easy to condense several hundred pages into a single thesis statement but I gave it my best shot. "Most studies have shown a link between people who eat more processed red meat and cancer. In particular, colorectal cancer."

"Most studies?" Katie asked.

Trying to summarize the entirety of the evidence was going to be next to impossible. It was hard enough to keep the attention of college students majoring in this stuff, let alone two random people you just met on the plane. There was a series of studies, going all the way back to 1959, called the Cancer Prevention Studies: CPS-I, CPS-II, and the CPS-II Nutrition cohort. Then there was the European Prospective Investigation into Cancer or EPIC study and the original Nurse's Health Study. These were all cohort studies, meaning that researchers recruited volunteers and periodically sent them questionnaires to collect information about their diet, personal history, habits, physical activity, and any medical diagnoses. They would follow them over time and record who got cancer and what type. Taken together, these datasets provided a wealth of data for researchers to analyze. The CPS-II alone recruited over a million people.

"IARC's report was just a summary of all these other studies?" Jim asked.

I nodded.

"So where's the controversy?"

"Everybody looked at the same data. But not everybody came to the same conclusion."

"Was there a Twitter war?"

"Mercifully this predated the modern phenomenon of Twitter wars. But after IARC published their report, the Nutritional Recommendations Consortium published their recommendations, which said that people shouldn't reduce their meat consumption and should keep doing what they're doing."

"I like how these people think," Jim said.

"I do not," said Katie.

"Well, if you went into this wanting to eat meat," I said, "then yes, the NutriRECS consortium's recommendations, as the group is called, came as good news that justified what you were already doing."

"I do like it when people validate things I'm doing anyway," Jim said.

"But it seems strange to me," Katie said, "that two groups could come to such different conclusions when looking at the same data."

"Well . . ."

"Wait a minute," Katie interrupted, "there wasn't some sort of conflict of interest here, was there? I mean nobody from that group got money from the meat industry or something did they?"

"Actually . . ."

"You're kidding?" Katie said.

I wasn't. A few months later it came out there were financial ties between NutriRECS and AgriLife Research, an arm of Texas A&M University that received funding from the beef industry. It was a messy revelation. There was never a suggestion their specific review had been paid for by any private entity, nor that the funding had influenced the outcome of the analysis. But it cast things in a negative light.

As much as Katie took issue with the perceived conflict of interest, Jim was struggling to reconcile how IARC and NutriRECS could have come to such diametrically opposed results. When I told him their reports were not that different when you examined them carefully, Jim said he had no idea what was going on anymore.

It was admittedly confusing. These two groups had reviewed the same studies and seen the same data. The difference was how they judged the quality of the evidence. IARC was happy with the quality of the research, while NutriRECS felt like they couldn't rely solely on observational studies and wanted randomized trials.

Once Jim and Katie were finished upbraiding me for my use of jargon, they gave me an opportunity to explain the difference. In observational studies, you let people eat whatever they want, and you simply observe them. As Jim put it, the voyeuristic approach to research. I had never heard it described like that before, and I was going to have to remember to use it during my next lecture. It would probably get a big laugh. The other type of research was experimental, where you randomly divide people into groups and tell one group to eat this and the other group to eat that to see if any differences emerge over time.

NutriRECS's preference for experimental studies is easily understandable. In observational research people choose how much meat they eat, if any. But people who don't eat meat are different in many ways from people who do. People who don't eat meat would likely be more health-conscious than meat eaters, for example.

"More ethical too," Katie added.

Jim had his counterargument ready. "But being vegetarian doesn't necessarily mean you're being healthy. A vegetarian can eat french fries all day, smoke, and eat tons of sugar."

Katie didn't seem like the type of person that smoked, ate french fries, or ate tons of sugar. She also didn't seem like the type of person who would resort to violence, which was good for Jim in this instance.

But his argument was a valid one. In observational research, people can be different in a multitude of ways. Vegetarians might be overall healthier than non-vegetarians, which would skew the data one way. But they might not and that would skew the data in the opposite direction. When you have multiple differences between groups, it becomes hard to tease out what effect any single variable has.

Jim complained that research was hard, which I admitted, and that I was making things unnecessarily confusing, which I denied. But since the whole issue seemed to hinge, at least from NutriRECS's point of view, on the randomized trials, Jim wanted to know what the randomized trials showed.

I had to tell him there weren't many of them. The biggest one to date was from the Women's Health Initiative. They did a very large study of almost 50,000 women. But they studied a low-fat diet, not specifically a meat-free diet. The women in the low-fat group did eat less red meat indirectly but that wasn't what they were looking at.

"Did the low-fat diet reduce the rates of colon cancer?" he asked.

"No, it didn't."

Jim smiled. If the data linking meat to increased rates of cancer wasn't as solid as it seemed, he could once again have bacon for breakfast with impunity. I felt bad dashing his hopes. "Well, there are some caveats."

"There are always caveats with you," he said.

It was surprising how often people told me that.

"The low-fat group reduced the amount of meat in their diet by a little bit, but not by that much. Maybe one to two servings less per week and that's probably not enough to make a difference."

"Then this study doesn't really move the needle, does it? It doesn't really add anything one way or the other."

"Not really," I agreed.

"Aren't we back where we started then?" Jim asked.

"Hold on." Katie jumped in. "Even if you don't have any randomized trial data there are still all those other studies you mentioned. All the observational type studies from all those different groups that IARC reviewed."

"Ah, but in those studies, people decided by themselves what they were going to eat," Jim shot back. "So maybe the people who ate less meat were different for other reasons, right?" Jim looked to me for validation.

"Maybe," Katie answered back. "But there were multiple studies, by different groups, over many years, involving millions of people. Maybe one study could be skewed one way or the other. But how could you keep getting the same result, by different groups, consistently, in the same direction, multiple times, over so many years?"

Jim, though he had come a long way as an amateur epidemiologist since we had met at the airport, didn't have a ready answer. I answered for him. "I think what you two are getting at is the fundamental problem with this question. Ideally, you would want to have randomized trial data to say definitively if eating red meat and processed red meat causes cancer. But we don't have that. So the question becomes: Do you feel comfortable relying on the large amount of observational data, imperfect though it is, and recommend people eat less meat?"

"No," Jim said.

"Yes," Katie said.

They looked at each other and I expected one of them to say *jinx* but neither did.

Katie looked at me. "Can't groups just get their act together? It's not helpful if different people look at the same evidence . . . wait. Just to be clear, the IARC and NutriRECS group didn't disagree about the actual data, right? Just the interpretation and the quality of the data."

"Pretty much."

"Okay, so can't they just sort out their differences? It's not helpful to have this type of back and forth. People are going to start to believe that nobody knows what they're talking about."

"I think they already believe it." I found a screenshot of a cartoon I keep on my phone. I often use it in presentations to my students because I think it perfectly encapsulates how some people think about medical news.

"That seems accurate," Jim said.

"It's satire but not by much. When you keep flip-flopping back and forth and say meat is bad one day and meat is okay the next, then people won't know what to believe and will eventually stop listening."

"Does this happen often?" Jim asked.

"It does."

"Anything you can do to stop it?"

I almost laughed. "Me personally? No, I don't have that much authority." What could I possibly say? "Not every study is well done. When it comes to food, and especially when it comes to cancer, there's a lot of flip-flopping."

"If it's just a question of interpretation, then I'm probably not going to stop eating meat."

Katie sighed. "Why, Jim? Why?"

"If a group of experts like the NutriRECS group are going to come out and proclaim —"

"Proclaim might be putting it strongly."

"They all agreed —"

"Of the 14 members on the panel, only 11 voted for the recommendation to continue eating red meat. Three of them voted for a recommendation that people should reduce red meat consumption."

"Ha!" Katie was pleased by that.

"I'll concede it's weird that it wasn't unanimous," Jim said. "But 11 of those 14 people voted to strongly recommend . . ."

"It was a weak recommendation. A weak recommendation based on low-quality evidence."

"Ha!" Katie said.

"I'm confused," Jim said. "That recommendation is much less reassuring than it was 30 seconds ago."

"It would appear," Katie said, "they told people they could eat meat, but they didn't feel that strongly about it."

"If there was no consensus, why didn't they say nothing?" Jim asked.

"My experience has been that academics are too enthralled by the sound of their own voice to remain silent on an issue." Katie looked at me. "No offense."

"Argue as much as you like," Jim said. "But if there was no observable benefit to cutting out meat from your diet —"

"There was a benefit, just a very small one," I said.

"How small?"

"They estimated if people cut out three servings of meat per week from their diet, you would prevent seven cancer deaths per one thousand people."

"That seems like a pretty small number. That's just . . . ," Jim closed his eyes, "0.07 percent less, right?"

Katie rubbed her temples. "0.7 percent."

"Seven fewer cancer deaths per one thousand people may not sound like a lot. But you multiply that out by millions of people in this country, and that's a lot of cancer. It's definitely not negligible."

"It would appear, even the people telling you to keep eating meat say that it increases the risk of cancer," Katie said.

Jim swallowed hard. "Is there a way for me to know what my risk of cancer is?"

"If you go back to the IARC report, the average lifetime risk of colon cancer is about 4 percent. Eat 50 grams of processed red meat every day and that goes up to 5 percent."

Jim exhaled. "That's not too bad. Pretty minor if you eat meat every day, which I don't. Not every day. I was expecting worse."

"We call that the population paradox. Something may have a small benefit for you as an individual, but when you apply it to the population as a whole, the benefits add up in significant ways."

"You see, Jim," Katie chimed in, "if everyone stopped eating meat, the world would be a better place and everyone would be healthier."

"I wouldn't be much healthier in the grand scheme of things. I don't know. I feel like it's not the most important thing I need to worry about. I mean, it's not as bad as smoking, right?"

"Smoking is far worse," I said. "Eating processed red meat isn't good for you, but it's clearly not as bad as smoking, especially if you don't eat meat every day."

"I'm just going to remind you that you could not eat meat at all and be even healthier," Katie said.

"Maybe I could cut back a bit," Jim said.

Katie seemed to be having an effect on him. I could only hazard a guess as to why.

He started poking at his meal again. "Either way, I'm not eating this, whatever it is."

"I still think it's couscous," I said.

"I don't know . . ." Jim said and contemplated his meal. "What I am certain about is that I have some granola bars in my carry-on."

Katie and I were also very certain that that sounded like a good idea.

MYTH #3

Some Salt Is Good for You

Even though I can't usually sleep on planes, I closed my eyes and tried to doze for a bit as Jim and Katie talked. Katie tried to convince Jim about the merits of vegetarianism, then Jim started talking about his career as a loss adjuster. Sensing that he was losing his audience, he switched to fishing, which just made matters worse. He recovered a bit when he talked about his favorite fishing locales (Mexico, Costa Rica, Hawaii). Soon he and Katie started comparing travel lists, which allowed him to de-emphasize their obvious differences.

About halfway through Jim's story about his trip to Colorado, the flight attendants came by to offer up our after-dinner snack. The choices were sweet or salty, which amounted to a choice between a cookie or a bag of pretzels. I, not particularly caring either way, chose the cookie, as did Katie. Jim, however, turned out to be something of a cookie snob and chose the pretzels. He had some very strong opinions on what made a good cookie. Katie, who it turned out was one of those people who bakes to relax, had just as much to say on the subject.

The debate ended up being a wide-ranging affair that jumped from topic to topic with almost no segue. Should cookies have fruity jam filling? Was fruit better than chocolate? Should the texture be sandy or more cakey? Dipped or chipped? Was shortening a valid substitute for butter? Jim hated shortening for its waxy aftertaste and cited a particularly unsatisfying Christmas cookie experience, whereas Katie argued that as a butter substitute, it was cheaper and a plant-based alternative for people who, like her, didn't feel that animals should suffer for our pleasure. The debate on shortening and its societal, environmental, and social implications quickly transitioned into a more general debate of soft versus hard cookies. Jim preferred soft chewy cookies, whereas Katie proclaimed hard cookies, and particularly biscotti, to be the pinnacle of baking skill.

"You see," she explained, "anyone can just plop a scoop of cookie dough on a baking pan and say they made cookies. But biscotti take skill."

Katie, it turns out, was half-Italian. Biscotti were a source of pride for her, and she defended her position without backing down. It turns out that when you make biscotti, you have to bake them till they're nearly done, then take them out of the oven, cut them up and then bake them again. That's why they're called *biscotti.* It means "baked twice" in Italian.

She and Jim went on like this for some time and only ever agreed on one issue. At one point, I tried to join the conversation by telling them that, chemically speaking, there wasn't really much difference between artificial and natural vanilla. At which point, they both turned on me and said that anyone who used artificial vanilla in their baking didn't deserve to eat cookies. In the end, there's nothing like a common enemy to bring people together.

I ate my cookie in silence and had no major complaints. Jim's pretzels were still unopened, and I wondered if there was some graceful way I could suggest that he pass me his snack if he wasn't planning to eat it.

He eventually tore open the bag as he spoke with Katie and popped a pretzel in his mouth. He pronounced himself happy with his choice of snack. "You can't get good cookies on a plane," he said. "Cookies aren't meant to be wrapped in plastic."

To be fair nothing was really meant to be wrapped in plastic, but I didn't feel like pointing that out.

"They're very salty though," Katie said.

Jim shrugged. "Salt doesn't bother me."

"You aren't worried about blood pressure?" Katie asked.

"No, why?"

"Salt is bad for your blood pressure."

"I thought that was a myth, I read an article in *Scientific American* or the *New York Times* or somewhere that said it wasn't that bad."

It was actually *Scientific American*.

SCIENTIFIC AMERICAN

THE SCIENCES

It's Time to End the War on Salt

The zealous drive by politicians to limit our salt intake has little basis in science

By Melinda Wenner Moyer on July 8, 2011

The article highlighted some recent studies that cast doubt on the benefits of low-salt diets and made the case that our focus on removing salt from our diet was ill-advised. We would be left with bland french fries and not be any healthier, so why not just leave things be? It was a seductive message. I remembered reading it at the time and thinking it would be nice if things were that simple.

Jim and Katie were still arguing about the purported dangers of salt and blood pressure so I figured I should step in and explain away the controversy. "This whole thing is very messy."

Jim handed me his napkin.

I bit my tongue. "I meant the scientific debate around salt has gotten very messy. The problem is information bias."

"You mean racism?" Jim was confused. "I'm against racism," he said definitively.

He didn't follow up the claim with any elaboration, but I figured that his heart was in the right place. In this day and age, that was a good start. I explained to Jim and Katie that information bias didn't have anything to do with racism or discrimination.

In statistics, the term *bias* means something that can skew the results of your research. There are many different types of bias, but *information bias* refers to errors in how we collect and measure data. For example, if you did a survey and you underestimated how many people in the population smoke, for whatever reason, that would be a potential source of error and information bias.

"Can't researchers just double-check the data? My company gets outside auditors all the time for exactly that reason," Jim said.

"That wouldn't work," I said. "Because" — I struggled to explain this clearly — "it also depends on whether the error is random is not."

"Are we going to talk about vitamin C again?" Jim asked.

I told him we weren't. Katie raised an eyebrow at the non sequitur, and I told her the abbreviated version of the story of how Jim and I met. It sounded slightly implausible when I said it out loud, but life is implausible if you look at it objectively.

I sipped my water to stall for time and decided that I'd just have to go for broke and explain why information bias can be such a problem in medical research.

"Let me try to explain it to you this way. Do you know how much you weigh?"

"Not really," Jim said.

Katie was shocked. "You don't know how much you weigh?"

"Nope. I think around 160 pounds, but I haven't weighed myself in a while." Jim shrugged.

Katie looked to me for validation. I just shook my head too. "Unbelievable," she said.

"I don't really care enough to check," he said. "Plus, I don't own a scale, but I think I have a rough estimate."

"I am truly fascinated by the fact that you don't know how much you weigh," she said.

"It's a gift."

"Probably a guy thing," she mumbled under her breath.

I didn't know how much I weighed either. Nor did I own a bathroom scale, so Katie was probably right. It probably was a guy thing.

But even if somebody did weigh themselves regularly, the human brain does something interesting when it processes numbers. People are either 165 pounds or 170 pounds. Nobody says they're 168 pounds. We like round numbers, and without realizing it we round things to the nearest tens or the nearest fives, depending on the scenario. In most situations, this doesn't matter. But when you're analyzing data in medical research, precision is important.

Amazingly, this type of error may not matter as much because it tends to balance out. If one person is 172 pounds and another person is 168 pounds and they both fill out 170 pounds on their medical questionnaire, the final result may balance out. The problem is when things skew consistently in one direction over another. Because people consistently overreport their height and underreport their weight.

"Unsurprising," Katie said. "I'd be shocked if people claimed to weigh *more* than they actually did."

"Agreed. But either way, the point is you have to adjust for this discrepancy statistically."

"Or just weigh and measure people directly," she said.

"Or that, yes."

"So people lie, is what you're saying."

"People aren't lying exactly. At least not consciously. Information bias in research isn't about deception. It's a complicated concept. There are something like 78 different types of biases described in the literature."

"Hold on, let me make a list," Katie said. She neglected, however, to get out pen and paper.

I pressed ahead. "The first problem is just poor recall. If people don't weigh themselves, they are basically doing nothing more than guessing. And that's clearly inaccurate."

"Obviously," Katie said. She looked at Jim. He continued to eat his pretzels.

"But the other issue is something called social desirability bias. That's when people answer questions in a way that de-emphasizes negative behaviors."

"Sounds like online dating."

"I wouldn't know. I never got into that."

"You're not missing anything," Katie said. "Online dating is where truth and self-awareness go to die."

I tried to change the subject. "Social desirability bias happens when you're doing a survey and people consciously or subconsciously tell you what they think you want to hear."

"So exactly like online dating then?"

I wasn't going to take the bait. I did have an online dating account once. My friends set it up for me, but after a month I lost faith in humanity and deleted it. My first match was a woman whose profile picture was her licking a butcher's knife. Never again.

Both Jim and Katie had their own horror stories about online dating, and it was some time before either of them wondered how we'd gotten onto this subject. They not unreasonably wanted to know what any of this had to do with online dating.

I admit it was a bit of a subtle concept. But social desirability bias happens because people think there's a "right" or a "better" answer to a question. People will say they're taller than they are, or that they weigh less or smoke less or exercise more.

But the issue is more complicated with food. Say you were doing a study on apples, and you invited people to come in and take a survey about their health, and you specifically asked them how many apples they ate on a weekly basis. They would probably assume that you had a special interest in apples because they were healthy and did something to prevent disease. They would then tend to round up because they wanted to give you the "right" answer. Peer pressure doesn't end in high school.

"Can't you just tell people to be honest?" Jim asked.

"That's kind of like telling someone 'Don't look behind you.' It has the opposite effect. You say that and they always look behind them."

"Is this an issue with all food questionnaires?" Jim asked.

I shrugged. "It's almost inevitable when you ask people to remember stuff that happened way in the past."

"Because memory is unreliable?"

"Very. It's hard to get people to report their eating habits from six months ago. People can't accurately remember that far back. It gets even more complicated when you introduce concepts like recall bias."

"Is this one of your 78 bias types?" Katie asked. I wasn't sure if she was being sarcastic. She was hard to read sometimes.

"I can probably explain that better by continuing the apple analogy."

Jim felt compelled to make a joke about how an apple a day keeps the doctor away.

With that out of the way, the road was clear for me to explain the issue of recall bias. It happens when the presence or absence of a disease affects how people remember past exposures. Imagine if researchers wanted to know if apples affected stroke risk. Participants who had a stroke in the past would be more likely to think hard when asked about their diet. Those who never had a stroke wouldn't. In that scenario, people without strokes would underreport apple consumption and people with a prior stroke would overreport it. Ignore the phenomenon and you might mistakenly think apples cause strokes when in reality the two groups just remembered the past differently.

"I understand memory is a complicated thing," Katie said, "but is there something about food that makes it just that much more complicated?"

"Sort of. The thing about food is what you eat changes over time."

"Not me, I've had the same thing for breakfast every day since I was a kid," Jim said.

"Well, that's unusual in that . . ."

"Toast and jam," he added.

"Most people aren't that consistent, and what they eat changes over time. Food isn't always in season. You'll mainly have apples in the fall and strawberries in the summer."

"Greenhouses exist," Jim said.

I suppose Jim was right after a fashion. Nowadays, you can pretty much get anything all the time if you're willing to pay a premium.

But there's still some seasonality to what you can find in the grocery store, and people's eating habits are not generally as consistent as Jim's seemed to be.

Suffice it to say, recording such things accurately for research purposes is not easy. The obvious strategy is to have people keep track of what they eat in a food diary. And while that sounds straightforward, it quickly becomes tiresome for the people who have to do it every day. The longer the study, the less likely it is that people will stick with the diary.

The alternative most researchers go with is a food questionnaire, which is exactly what it sounds like. You periodically send people a questionnaire asking them about their food intake. Jim was curious about what it looked like, so I called up a copy of one I had on my laptop.

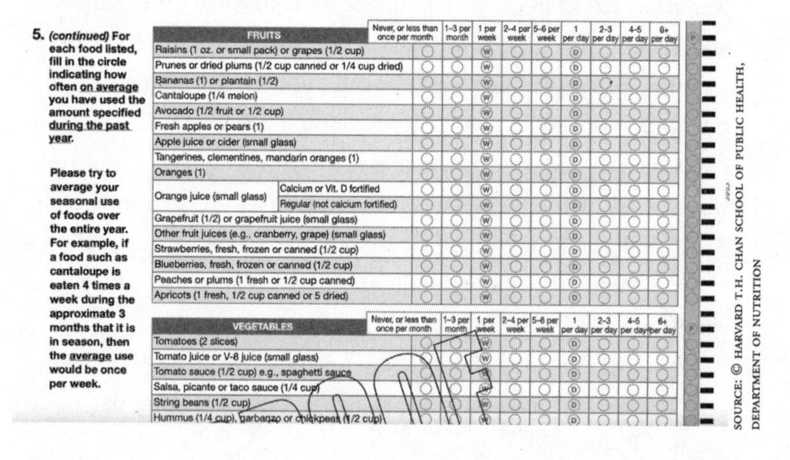

5. ***(continued)*** **For each food listed, fill in the circle indicating how often on average you have used the amount specified during the past year.**

Please try to average your seasonal use of foods over the entire year. For example, if a food such as cantaloupe is eaten 4 times a week during the approximate 3 months that it is in season, then the average use would be once per week.

FRUITS	Never, or less than once per month	1–3 per month	1 per week	2–4 per week	5–6 per week	1 per day	2–3 per day	4–5 per day	6+ per day
Raisins (1 oz. or small pack) or grapes (1/2 cup)			W			D			
Prunes or dried plums (1/2 cup canned or 1/4 cup dried)			W			D			
Bananas (1) or plantain (1/2)			W			D			
Cantaloupe (1/4 melon)			W			D			
Avocado (1/2 fruit or 1/2 cup)			W			D			
Fresh apples or pears (1)			W			D			
Apple juice or cider (small glass)			W			D			
Tangerines, clementines, mandarin oranges (1)			W			D			
Oranges (1)			W			D			
Orange juice (small glass) — Calcium or Vit. D fortified			W			D			
Orange juice (small glass) — Regular (not calcium fortified)			W			D			
Grapefruit (1/2) or grapefruit juice (small glass)			W			D			
Other fruit juices (e.g., cranberry, grape) (small glass)			W			D			
Strawberries, fresh, frozen or canned (1/2 cup)			W			D			
Blueberries, fresh, frozen or canned (1/2 cup)			W			D			
Peaches or plums (1 fresh or 1/2 cup canned)			W			D			
Apricots (1 fresh, 1/2 cup canned or 5 dried)			W			D			

VEGETABLES	Never, or less than once per month	1–3 per month	1 per week	2–4 per week	5–6 per week	1 per day	2–3 per day	4–5 per day	6+ per day
Tomatoes (2 slices)			W			D			
Tomato juice or V-8 juice (small glass)			W			D			
Tomato sauce (1/2 cup) e.g., spaghetti sauce			W			D			
Salsa, picante or taco sauce (1/4 cup)			W			D			
String beans (1/2 cup)			W			D			
Hummus (1/4 cup), garbanzo or chickpeas (1/2 cup)			W			D			

SOURCE: © HARVARD T.H. CHAN SCHOOL OF PUBLIC HEALTH, DEPARTMENT OF NUTRITION

Jim approved of the multiple-choice nature of the questionnaire because he had serious reservations that anyone would choose to write out their answers longhand. But he took issue with the instructions for filling out the form.

> Please try to average your seasonal use of foods over the entire year. For example, if a food such as cantaloupe is eaten 4 times a week during the approximate 3 months

that it is in season, then the average use would be once per week.

He looked at me and cocked one eyebrow. "I shouldn't need an advanced math degree to calculate how much cantaloupe I eat in a year."

You didn't exactly need an advanced math degree, but I agreed it was pretty complicated.

Jim was scrolling through the file to see how long the questionnaire was. It was clearly longer than expected. "How long does it take to fill in one of these?"

"Average is somewhere between 30 and 60 minutes."

"You realize that most people lose interest if something takes more than five minutes, right?" Katie asked.

She had a point. Filling out questionnaires can be tiresome. At medical conferences, we now have to fill out evaluations after every speaker. I confess I don't always give these reviews my full attention, and to get through them I sometimes just click on "Meets Expectations" for everything, and I call it a day. I've always wondered if people who fill out food questionnaires do the same. I decided I didn't want to know the answer.

Jim had a slightly more optimistic view. He argued that if someone was going to bother volunteering for a research study in the first place, they wouldn't do so just to blow through the questionnaires.

"Your faith in humanity is charming, if possibly misplaced," Katie said.

"Another problem," I said, "is that people don't always prepare their own food. People eat out a lot."

"I've discovered that most unmarried males of the species have nothing but a lightbulb and an expired box of baking soda in their fridges," Katie said.

I offered up no rebuttal, but Jim felt the need to protest. To avoid getting back into another argument about cilantro or some other topic foodies care about intensely, I tried to veer the conversation back to more familiar ground. "Katie is unfortunately correct," I said.

"People eat out a lot or buy ready-made food. That's not a problem per se but it does mean they don't directly know what's in their food."

"You mean like all the added sugar that's in stuff, and we don't realize it?" she asked.

"That too. But take something more basic. Eggs. If I asked you how many eggs you eat in an average week, what would you answer?"

"Let me think," Jim said. He was silent for a moment. "Well, I don't eat eggs every day."

"But what about eggs in baked goods?" Katie said.

"I forgot about that. Does that count?" Jim asked.

"You should be counting all your eggs," I told him.

"Before they hatch?" Jim waited for us to laugh.

"This might be useful," I said to spare him an awkward silence. I called up a cartoon we'd made to illustrate the problem to students.

CARTOONIST: LUC SPERL

It's an inherent problem with food questionnaires that researchers look to them for a precision that human memory simply can't provide.

"So researchers are deluding themselves," Jim said.

"I wouldn't go that far," I said. "But we have to recognize food frequency questionnaires are not perfect, especially when you ask people to average out their eating habits over long periods of time."

"You're asking them to guess."

"We tell people not to, but I'm afraid many probably do."

"Is it more accurate for some foods versus others?" Katie asked.

"Salt is tricky because so much of it in our food is unseen."

"Why?"

"All food has some sodium in it. When we say *salt* what we usually mean is table salt or sodium chloride."

"No chemistry," Jim said.

I promised to be more circumspect going forward. But I did have to stress that we needed to talk about sodium intake, not salt intake, because you can get sodium from things that are not table salt. About 5 percent of the sodium we consume is found naturally in food. Another 15 percent comes from salt we add to food during cooking. But the remaining 80 percent is added to our food by other people.

"I'm pretty sure people aren't breaking into my house and pouring salt on my vegetables," Jim said.

"Perhaps not but most of the salt you eat is added during food manufacturing. Processed foods, junk food, pre-packaged stuff, take-out, restaurant food, all that stuff has a lot of salt in it, and it's largely outside of your control."

"You could just not eat it," Katie said.

"That's best," I said. "If you cook at home, you invariably put less salt in your food."

Jim sat up straighter and proudly announced that he usually cooked at home. "I guess I'm doing everything right," he said.

"You still eat animals," Katie said, and Jim deflated.

Apart from the health implications of all that extra salt, the added salt in pre-prepared foods is a major problem for researchers who want to study the issue because people simply can't gauge how much salt they eat. While it's fairly straightforward to count the number of pears you ate in the past week, measuring how many milligrams of salt you ate is simply not feasible unless you measure out and cook all your food yourself. Even then, you're ignoring all the sodium that is naturally present in foods.

"What's the answer?" Katie asked.

"You can't really rely on questionnaires for salt. You have to measure it."

"Like in your blood?"

"No, the urine."

Katie was quiet for a long moment. "Why . . . urine?"

It might seem counterintuitive, but blood analysis won't tell you anything useful in these settings. Urinalysis will. If you can measure how much sodium was excreted in your urine over 24 hours, you get a pretty good idea of how much sodium you ate in that time.

Katie blinked. "But how do . . ."

"You have to get people to collect their urine."

"People do this voluntarily?"

"It takes some convincing sometimes, but people do it."

"I have trouble believing somebody would want to collect their own urine for a whole 24-hour period."

"There are people who do that sort of thing," Jim said.

Neither Katie nor I said anything at first. She proved to be the braver of us and asked, "Why do you have first-hand experience with this?"

"You meet weird people when you do online dating."

"You need higher standards," Katie said.

"You don't always learn what you need to learn from an online dating profile." Jim shrugged. "And sometimes you just have to take what you can get."

Jim didn't elaborate, and I was loath to press the point. Katie, perhaps to boost his morale, said he had a number of things going for him. He had a job, he knew how to cook, he still had his hair, and he was able to board a plane, which meant he didn't have a criminal record. Not many people on Tinder can lay claim to all those achievements.

That seemed like a low bar, but I don't date often enough to have an opinion.

All the talk about urine collection triggered something in my brain, and I eased out of my seat to use the bathroom. As Katie tried to restore Jim's self-esteem, I made my way down the aisle to the lavatory. The less said about the experience the better. When I returned to my seat, the pep talk was over and, I suppose to change the subject, Katie asked a few more questions about this novel idea of 24-hour urine collections.

"Do you really just tell people to pee in a cup all day?"

"It has to be 24 hours' worth of urine, so it's more of a jug. But, yes, that's it."

"Won't it be thrown off by how much water you drink?"

"Water doesn't affect it. You'll produce more urine if you drink more water, but the amount of sodium in the urine stays the same."

With a few exceptions, the amount of sodium in your urine really does reflect how much sodium you ate. It fluctuates based on how much sodium you ate on that day, so the ideal would be to get multiple 24-hour measurements. That way, you can average them out and eliminate the outliers. The limitation is the burden you put on patients, who have to carry the jug with them everywhere they go.

"Wait, so you have to take it with you to work?" Jim asked.

Invariably you do.

"What if somebody misses a collection?" Katie asked. "What if they go to work and forget to bring the collecting jar with them and have no choice but to just go to the washroom as they normally would? Won't that throw off the sample?"

"It would," I said. "No research tool is perfect."

I think it would surprise most people to learn how crucial seemingly mundane decisions are in medical research. Choosing how you're going to measure sodium intake wouldn't strike most people as a particularly complicated question. But every decision has its advantages and drawbacks.

A food questionnaire is obviously easier to administer and less expensive, but you sacrifice accuracy. Urine collections are more accurate but more time-consuming and difficult for people to follow through with. There are also different types of urine collections. You can do a single or "spot" urine collection. Those are easier for people because they only have to collect a single urine specimen. But spot urine collections have greater variability, and some mathematical adjustment is needed to estimate the 24-hour sodium intake. A full 24-hour urine collection is more accurate but harder for patients to do. Multiple 24-hour collections are even better but incrementally harder for patients to comply with. There are always trade-offs between accuracy and simplicity.

"There must be an easier way to do this?" Jim asked.

"I'm all ears," I said.

Katie ventured a suggestion. "Couldn't you do a study where you take a bunch of people and divide them into two groups? One gets a low-salt diet and the other a high-salt diet?"

"What you're suggesting is called a research kitchen. That's where you make everybody's food for them. That would be the better way."

"So why doesn't everybody do that?" she asked.

"Money mostly. Not everybody has the research funds."

"But what if they did?"

"Even then, it only works if people don't get food from outside sources."

"That seems like it would be hard to control."

"You could do it in the short term, but it's practically impossible in a long-term study."

If you wanted to do a short-term study, you could theoretically have patients stay in specially designed lab spaces where you served them all their meals. That would guarantee that no outside food sources were skewing the results. But while that might be feasible for a few days, anything beyond that would be impossible. It takes years to develop high blood pressure, and you can't maintain a study like that for that long.

"I don't see how you could make it work," Jim said.

"Agreed. That's how we got into the whole 'experiment on prisoners' thing," I said.

I was distracted by the flight attendants' refilling people's water glasses, so didn't realize Jim and Katie were looking at me funny.

"I'm sorry, the what?" Jim asked.

"Have you two never heard of the Salt Wars?" I asked.

They looked at each other and then back at me. "You mean *Star Wars*?" Jim asked.

"Salt Wars. The salt controversy thing. A group of researchers got together and decided to experiment on prisoners? It was a pretty big deal."

"We clearly have different definitions of what constitutes a big deal," Katie said.

"Do you mean the Helsinki Declaration?" Jim asked.

"No, I was referring to . . . I better tell you the story."

"Short version," Jim said.

I did my best to oblige but there was a lot of backstory to the salt controversy. It had been clear for a while that eating a lot of salt was bad for you. Studies like the INTERSALT study found high-salt diets in populations correlated with high blood pressure. Then other studies like the DASH study and the TOHP study found that getting people to eat less salt prevented heart disease.

"So far so good. What's the problem?" Jim asked.

"The problem is what to put in official guidelines. When people ask, 'How much salt am I supposed to eat?' what are you going to tell them?"

"Less," Jim said.

"Much less," Katie said.

"But what if people want a specific amount or threshold?" I asked.

"Do you need to have a specific number?" Katie asked.

"People usually want one."

"Tell them to learn to live with disappointment," Katie said.

I don't know if Katie's bluntness would have served her well on a guideline committee. They do usually want specific thresholds or targets if only to guide public policy and treatment decisions. And for a long time, that number for salt was 2,300 milligrams of sodium per day. Different groups have issued different guidelines over the years, and some went as low as 2,000 milligrams per day, so there was some variability in the literature.

"I have a couple of questions," Katie said. "One, how did people get these numbers? Two, how come everybody came up with different numbers?"

Those were good questions that unfortunately did not have good answers. There was no firm data, and that was part of the problem. The source of the controversy was some people didn't think there was enough solid data to support the specific targets laid out in the guidelines. Some people were happy with the quality of the evidence and others wanted more.

"How do you resolve something like that?" Jim asked.

"The best way to settle this is to do what Katie said."

"I never said experiment on prisoners!"

"I meant you need a randomized trial."

But it wasn't simply a question of people debating the quality of the evidence, an interesting but largely theoretical argument, interesting to academics but no one else. Had the matter rested there, it would have been bad enough. The added complication was the publication of data in 2014 suggesting that current salt guidelines were setting a threshold that was too low and that reducing salt intake too much was dangerous.

"Where did that come from?" Jim asked.

"There was a research study called the PURE study, the Prospective Urban and Rural Epidemiological study —"

"Catchy name," Jim said.

Medical researchers do like their acronyms. But while the name was simple, the PURE study wasn't. It generated several publications with different goals and outcomes. The first was clear-cut. It looked at the association between salt and blood pressure and found a consistent straight-line relationship. There was a greater effect in people who had high blood pressure to begin with, but it held true for everyone, which made it more or less consistent with previous studies like INTERSALT.

But when researchers looked at cardiac outcomes, the results were less straightforward. They found what's called a *U-shaped association*. The people who ate very little salt did poorly, the people who ate a lot of salt did poorly, and the people in the middle had the lowest risk. I drew out what the curve looked like for Katie and Jim since these concepts can get a little abstract without visual aids.

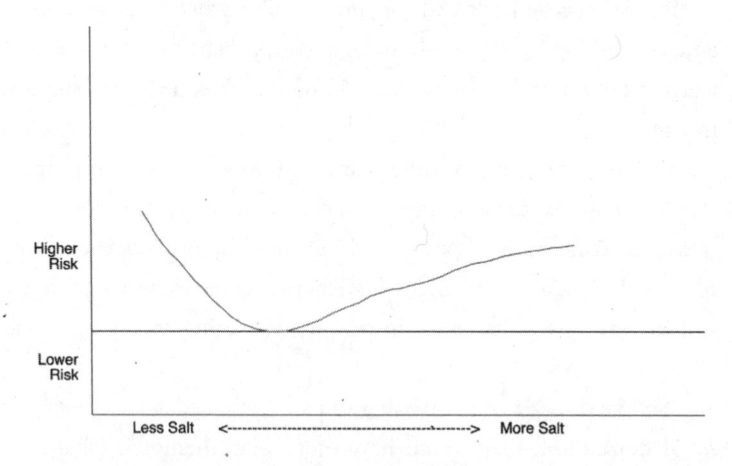

Jim interpreted this to mean "everything in moderation." People often use this saying to justify eating things that are not good for them. But the problem with the PURE results was not everybody agreed with them. "The issue was how they measured sodium," I said. "It brings us back to the issue of information bias."

"It took you an awful long time to get there," Katie said.

"We did waste a lot of time talking about online dating."

"You missed some of Jim's crazy dating stories when you went to the bathroom. They were definitely worth it."

"I'm intrigued."

"Jim actually went out with the girl whose profile picture was her licking a butcher's knife. Guess how long they dated for?"

"Three dates."

"Six months."

"Impressive." I would have thought three dates more than enough to give someone the benefit of the doubt before running for the hills.

"I'd like to hear more about measuring salt in people's urine," Jim said.

"I compliment you on your commitment to epidemiology," I said.

"I just wanted to change the subject."

"I know."

Partly because I felt bad for Jim and partly because I genuinely enjoyed talking to people about this stuff, I went through the critical issue with the PURE study: How did they measure sodium intake?

The answer was a spot urine sample. That is, a single urine specimen and not a 24-hour urine collection. This became the main source of contention. Was a single urine sample good enough to address the question at hand? The debate became increasingly nasty and increasingly polarized with people associating themselves with different camps.

"So like pro-salt and anti-salt groups?" Jim asked.

"I don't think they would have ever called themselves that, but essentially yes. That's what it came down to."

"Who was right?"

History will probably tell us that no one was right. Supporters argued that the PURE study was the largest study of its kind in recent memory and collected data on over 150,000 people in 17 different countries. Getting 24-hour urine collections from all those people would have been complicated, if not impossible. It became a question of not letting perfect become the enemy of good. But the counterargument was: if you're going to start claiming low-salt diets are dangerous, then you have to back it up with solid data. Detractors didn't think spot urine specimens were good enough. That was the crux of the debate.

"So how did they work out their differences?" Jim asked.

"They didn't. It got ugly. People accused each other of being bought off by the food industry. People fought over who would sit on the guideline committees. It got a little bit out of hand."

Katie put her hand on my shoulder. "I'm struggling to understand how we went from a bunch of academics arguing about the best way to collect other people's urine to 'let's experiment on prisoners.' I feel like we skipped a few steps."

There actually weren't that many intervening steps. The two sides became progressively more and more estranged. The people who thought salt was bad weren't swayed by the other side's data and vice versa. And after years and countless papers, editorials, accusations, and name-calling, in 2018 they decided to have a meeting and settle the issue once and for all. Both Katie and Jim agreed that this seemed like a sensible course of action. After all, one expects eminent researchers to behave like adults and not squabble interminably like children.

"This was a large academic conference of researchers from across the planet?" Katie asked.

"No, it was six guys meeting privately in Jackson, Mississippi," I said.

"And these people . . ."

"The Jackson 6."

"Is that literally what they called themselves?" Katie asked.

"It is."

"Did they call themselves that ironically or . . . ?"

"I don't think it was official. David McCarron, one of the original six, came up with the name. But they were later joined by two other people including former, and now once again current, FDA commissioner Robert Califf, so in reality, there were eight of them in the end."

"All men?" Katie asked.

"Ummm . . . yes."

"I see. Figures. Proceed."

"The idea was to get six people together, three from each side, to hammer out an agreement. They added the extra two 'neutral' people to make eight. Then they published an editorial in the journal *Hypertension*, which laid out their plan."

"The plan being to experiment on prisoners?" Katie asked.

As hard as it is to believe, that is exactly what they suggested with seemingly little insight into what the public reaction would be. Their reasoning went something like this. They realized the situation had degenerated to such a degree that the only way to settle the issue was to do a randomized trial in a setting where you controlled people's food. The Institute of Medicine had suggested exactly that back in 2013. They came up with three possible venues: prisons, nursing homes, or the army.

"I say army," Jim said.

"None of the above! For funk's sake Jim!" She didn't actually say *funk*.

"Sorry, I change my vote," he said.

"We're not actually voting," I said.

"I feel like they needed to brainstorm more ideas," Katie said.

That was a common reaction at the time. But coming up with a working idea isn't easy. Even those ideas didn't stand up to closer examination.

"They ended up scrapping the army idea early on," I said.

"For ethical reasons?" Katie asked.

"Because soldiers tend to be young and healthy. If your goal is to study heart disease, you don't study it in healthy 20-year-old men, regardless of how much salt you feed them."

"Their concern was that not enough of their research subjects were going to get sick and die?"

Put like that, it sounded bad. But given the average age of most tenured professors in academia, it was highly likely the investigators would develop heart disease before the study recruits. That's not generally a recipe for success.

Jim nodded in agreement. "The nursing home setting would solve the age problem."

"Except many of them probably already had heart disease or strokes or a host of underlying medical problems," I said.

"Very true," Jim said. "That would mess up the statistical analysis probably." He looked over at Katie, who was frowning. "Plus . . . we . . . shouldn't experiment on the most vulnerable members of our society."

Credit to Jim. He recovered quickly on that one.

"That left only one option. Prisons. Not too young, not too old, and a setting where you can control what people eat."

"I have a question," Katie said. "At any point in these discussions, did anyone point out that experimenting on prisoners is something Nazis would do?"

This was going to be tough to explain. "I don't want to deflect the question."

"Then don't."

"Even the United States experimented on prisoners back then."

Most people have never heard of the Stateville Penitentiary Malaria Study, but experimenting on prisoners and purposely infecting them with malaria was not only tolerated but also celebrated. In 1945, *Life* magazine had a photo story titled, "Prison Malaria: Convicts Expose Themselves to Disease So Doctors Can Study It." Researchers from the University of Chicago ran it, and the U.S. Army sponsored it. Fighting malaria was important to the war effort, which is why both the U.S. and Germany were studying it. When Nazi doctors were put on trial in Nuremberg for their prison experiments, they claimed such experiments were common practice and pointed to the Stateville Penitentiary Malaria Study as proof. Prior to World War II, there were no international standards for what constituted ethical and legal experiments on human beings. That deficit became a problem at the Doctors' Trial at Nuremberg and led to the development of the Nuremberg Code.

"The more I learn about history, the less I want to learn about history," Jim said.

"I'm not going to lie to you. There's a lot of moral gray area in this story."

"So isn't experimenting on prisoners banned? I thought that was the point of the Nuremberg Code and the . . ."

"Helsinki Declaration," Katie said.

"It's not banned, but there are limits. The U.S. Department of Health and Human Services limits federal funding for prison studies unless you fall into one of five categories," I said. That usually means research that directly relates to prisons, criminal behavior, or diseases that specifically affect prison populations, like hepatitis or addiction. But research is also allowed if it directly benefits the prisoners, which is what the Jackson 6 argued.

"Do prisoners really care about low salt in their food?" Katie asked. "Don't they have more pressing concerns like not getting stabbed and finding work once they get released from prison?"

"I imagine those problems are more top of mind," I said.

"Prisoners get heart disease, like everybody else," Jim said.

"But it might not be top of the list of things they worry about. Not sure they care enough about the issue to volunteer for a study like this even if they were given an option," I said.

"What do you mean 'if'?" Katie asked.

That was another can of worms. The editorial explaining the plan to conduct the salt research didn't actually explain how the study would be conducted. That was going to have to be worked out later and was contingent on ethical approval from oversight bodies. But one possible design was a cluster randomized trial.

In a simple randomized trial, every person gets slotted into Group A or Group B by a coin flip or, in modern times, a random number generator. But it would be far easier to serve everyone in one prison salty meals and everyone in another prison low-salt meals. If you randomly assign prisons, not prisoners, to low- or high-salt diets, then you're doing a cluster randomized trial. That would solve many logistical problems but would essentially mean that the inmates wouldn't get to decide for themselves if they wanted to join the study. But this issue was largely theoretical since no research plan was ever tabled. The proposal unraveled too quickly.

"We don't know what the study would have looked like," I said to forestall any further objections. "The plan never really went anywhere."

"Was there a public backlash against this proposal?" Katie asked.

"Some. But the real problem was the funding."

"Please don't tell me the study was going to be paid for by the salt industry?"

Since she asked nicely, I didn't say anything.

"This is just like what happened with the meat study we were talking about before."

"There are a lot of parallels," I admitted.

"So this scandal killed the study?"

"It did."

In a story that already had many sharp turns, the final funding revelation highlighted just how strange the whole saga was. One of the coauthors had taken money from the Salt Institute, a lobby group for the salt industry, and not told his partners in the venture. The scandal derailed the plan and their grand announcement for a prisoner study was never put into practice.

"I'm glad this study fell apart," Katie said. "I am, as a general rule, opposed to doing things Nazis thought were a good idea."

Jim scratched his ear. "Me too. But serving people salty food isn't the worst thing we humans have ever done to one another."

"It's the principle of the thing," Katie said.

"I get it. But there should be a way to do this right. You could figure out consent and freedom to volunteer and all that."

Katie seemed unconvinced. "Hasn't anyone tried to do something maybe a little bit less . . . problematic?"

I scanned my memory. I don't pretend to know every study ever published, but I do my best to keep up with the medical literature. Some people are good at remembering sports statistics. I was built for this apparently. "There was the SSaSS study . . ."

"Wait, they actually called it the 'sass' study?" Jim asked.

"It was S-S-A-S-S. It's an acronym for the Salt Substitute and Stroke Study."

"And a salt substitute is something like sea salt or kosher salt?"

"Actually, those are all the same thing."

"No, they're not. When you're cooking, using sea salt instead of regular salt makes a difference. It changes the texture of the food. There's a very different mouthfeel," Jim said.

I didn't think *mouthfeel* was a real term, but Jim and Katie insisted that it was.

"It can really change the experience of the dish," she said.

"It might be coarser because they don't grind it down into fine granules, but chemically it's the same thing. It's still sodium chloride."

"We warned you about the chemistry," Jim said. Katie nodded.

"But what about pink Himalayan salt?" Jim asked. "That's obviously different."

It took a while to convince Jim that Himalayan salt didn't come from the Himalayan mountains. On the ground, we could have looked it up on Wikipedia and settled the issue, but on an airplane, I had a much tougher time. It was even tougher convincing him the pink tint of the salt didn't mean anything. Traces of magnesium, potassium, copper, and iron in the salt crystals give a pinkish hue and slightly different taste, but the crystals are still sodium chloride and are chemically identical to regular table salt. Jim put up some resistance. But after I emptied out the tiny packets of salt and pepper onto my food tray, the visual aids eventually carried the day.

"So there's no healthier version of salt then?" he asked.

"Not unless you get the salt with potassium instead of sodium in it."

"That exists?"

I told him it did. And the scientific data backs up its use. The SSaSS study I mentioned to Jim earlier tested a potassium salt substitute. They didn't use a pure potassium salt but rather a combo mixture that was 75 percent table salt and 25 percent potassium salt. But the fundamental point of the study was to use a low-sodium or lower-sodium alternative. And it worked. There were fewer strokes, less heart disease, and fewer cardiovascular deaths when you compared the salt substitute to regular salt.

On the surface, the SSaSS study should have settled the salt controversy since it showed quite convincingly that cutting sodium out of your diet prevented cardiovascular events. But there was some pushback from the skeptics. The study was done in rural China where the population tends to eat a lot of salt at baseline. So the study just proved that cutting back on sodium was good for you. It didn't establish what the lower limit of sodium intake should be.

"It's still a positive study," Jim said.

"But you have to weigh that against another recent study done in heart failure patients. It was called the SODIUM-HF trial."

"Do researchers just sit around trying to think of clever names for their trials?"

I didn't comment. But they totally do. This trial tested a dietary advice program to get people to eat less salt, and it made no difference. The main weakness of the study was advice can be ignored, and in fact, it was in this case. The group randomized to the new dietary counseling did eat a bit less salt than the control group but not by much. The difference was less than half a gram of sodium per day. So what the study actually proved was their dietary advice program wasn't very effective, not that aggressive sodium restriction is without merit.

"God, this is exhausting," Jim said as he flopped back in his seat. "It feels like every study has an asterisk next to it and we don't get any final definitive answer."

"Well, research is hard," I told him.

"I guess that's why they pay you the big bucks."

They really don't.

"I need a bathroom break," Jim said. "I'm going to ask for a drink on the way back. Anyone want anything?" Katie and I shook our heads. Jim got up and headed toward the rear of the plane. But Katie and I barely had time to say two words to each other before he returned. "That was disgusting. People are animals."

"Untrue. Animals are much less messy," Katie said.

Jim decided to change the subject. "Where did we leave things? Is salt bad for you, or is too little salt as bad as too much? I'm confused."

"It's complicated."

"I knew you were going to say that."

"For what it's worth," I said, "I'm pretty sure salt *is* bad for us. Nobody argues that a high-salt diet is good for you. We're just all debating what the lower limit should be. And no one is certain if that target should be under two grams per day or whether it can be higher."

"Except no one knows what two grams of sodium looks like," Jim said.

"Since most sodium comes from processed food, wouldn't it make more sense to just get people to eat less junk food?" Katie asked. "That's actually advice people could follow since two grams of sodium means nothing to somebody who ordered takeout."

Before I could agree, Jim jumped in to suggest that we should give people cooking lessons to wean them off the habit of ordering takeout. He was largely joking, but he had a fairly detailed plan about how most people could make dinner for themselves in under 30 minutes, which was largely how long it took takeout to arrive once you ordered it.

It was a compelling argument though I pointed out that most people lacked the necessary ingredients in their home. "I mean who actually randomly has cloves of fresh garlic in their kitchen?"

Jim and Katie looked at each other. "Everyone," they said in unison.

I doubted that was true, but Jim seemed certain that all self-respecting people had a robust collection in their kitchen of garlic, cinnamon, and parsley.

"Don't you use Italian parsley?" Katie asked.

"No . . ."

"Italian parsley has a more robust flavor, a better mouthfeel! Regular parsley is terrible. I mean maybe if you're making tabbouleh. But otherwise, no. Tabbouleh is the only acceptable use of regular parsley."

Jim capitulated, and Katie told him he needed to eat more veggie meals because, as she put it, "There's a better way to cook that doesn't involve killing Bambi."

Jim laughed, though I'm not sure he agreed. He said tabbouleh's main advantage was that it was easy to prepare.

"True, motivation is an issue," Katie said. "Sometimes it's hard to get motivated to cook if you're eating alone. I don't have anyone to cook for, apart from myself, so sometimes I don't bother making stuff that's really complicated."

Jim accepted this nugget of information without much of a visible reaction. I saw it, but I'm not sure Katie did. She tucked her legs under her and twisted so she faced Jim as they talked. I had little to contribute, so I sat back, closed my eyes, and let them have at it. At that point, I wasn't sure I could do much to help the process along. I figured it was largely up to them.

After we landed, I lost Jim and Katie in the scramble to exit the plane, but I caught up with them as we headed for the line of cabs. They were talking about a vegetarian restaurant in town and about whether vegetarian food designed to mimic meat actually tasted like meat or not. Jim was skeptical, Katie urged him to keep an open mind, and I tried to subtly suggest that an experiment was the only way to settle the issue. Having gained a newfound appreciation for the rigors of scientific experimentation, they liked the idea. They promised to be in touch, and we exchanged business cards.

As Katie went out to flag down a cab, Jim turned to me. "Me and Katie are going to split a cab because we're heading in the same direction, okay?"

I assured him I had no objections. "So is Katie going to get you to try a vegetarian restaurant?" I asked.

"Maybe," he said. "Maybe. I do really like meat though . . ."

"You can learn to like salad."

"That's true. Cobb salad is good."

"Jim, that has bacon in it."

"I know. It's amazing."

Even after this many years, every time I tell this story, that line gets a laugh.

"No, no, Jim," I rubbed my temples. "I was trying to make a point about . . . like the salad is an analogy for . . . you know what? Never mind. Just go for dinner and remember you don't always have to eat the same thing as the person you're with. People can be different and still spend time together. Just don't make a big deal about it."

"That's smart. Okay!"

"And just be yourself . . . or maybe a slightly idealized version of yourself."

Katie had secured a cab and waved him over. "I'll let you know how it goes," he said.

"I look forward to hearing about it," I told him. He smiled, clasped my shoulder, grabbed his bag, got into the cab with Katie, and they headed off.

MYTH #4

Coffee Causes Cancer

Once I got my cab, I headed off to my hotel. I planned to shower and change and, time permitting, take a quick nap before checking in to the conference. But when some hotels say they have a 3 p.m. check-in, they actually mean a 3 p.m. check-in. So with the freedom of someone who has the whole day ahead of him but the limitations of someone dragging a suitcase everywhere he goes, I headed to the coffee shop next door.

Small and independent, it had artwork on the wall that I neither understood nor recognized. But it was empty and had large couches that looked soft and comfortable and fit well with my plan to stretch out after the plane ride.

The only other person in the shop looked up from her phone, smiled, and said hi. "Welcome to the Golden Lion!"

She was incredibly enthusiastic for this early in the morning. "Hi there," I said. I stood my carry-on upright on its wheels and put my shoulder bag on an empty stool while I sat on the one next to it. I ordered a black coffee.

She giggled. "Sure thing!" She poured some into a mug.

I checked the name tag on her green apron. "Thank you . . . Casey." It wasn't a name you heard often anymore, but it matched her reddish-brown hair and freckled nose. Apparently, not everyone in California is a tanned surfer.

"You're funny." She giggled again, though I was pretty sure I hadn't said anything particularly amusing. I must have seemed confused because she added, "Nobody orders black coffee here. People order cappuccinos or lattes or espressos, but I've never seen someone come in and just order a 'black coffee.'" She said the last two words in a deep baritone an octave lower than her normal voice.

"What's your coffee, then?" I asked.

She tried to hide a smile by pressing her lips together, but the corners of her mouth turned up almost involuntarily. She looked around, lowered her head so she was closer to me, and whispered, "Don't tell anyone, but I don't really drink coffee." She giggled again.

You don't have to like something to sell it effectively. Telemarketers prove that every day. But I was curious how someone who doesn't like coffee ended up working in a coffee shop, so I asked her why she didn't like it.

"I don't know," she said. "There's all kinds of things on the internet. First, the WHO says coffee causes cancer. Then, two years later a judge says coffee needs warning labels. But I don't know what happened with that because we never put up signs or anything in the store. So I don't know."

I remembered the 2018 court case she was talking about. "That's not exactly what happened."

"So, is it like fake news? That's everywhere. It's terrible."

I was about to say something, but Casey started scrolling on her phone.

"Wait, it's right here." She read out the 2016 *Time* headline. "The Problem with Your Coffee. Hot Drinks a Probable Cancer Cause, Says WHO."

I did my own quick search and found a different article, also from *Time*.

"How Coffee Can Help You Live Longer," Casey read out loud. "Well, now I'm confused!"

"The research on this has been a bit all over the place."

"Are you a doctor or something?" she asked.

TIME
@TIME

How coffee can help you live longer

time.com
How Coffee Can Help You Live Longer
New findings add to growing evidence that coffee may actually have some benefits

9:45 PM · Dec 25, 2016

SOURCE: TWITTER

TIME
@TIME

The problem with your coffee

time.com
Hot Drinks a Probable Cancer Cause, Says WHO
In its latest report, the World Health Organization(WHO) reclassifies hot beverages, which include coffee, tea and mate, as probable cancer risk for ...

7:00 AM · Jan 8, 2017

SOURCE: TWITTER

"I'm here for a conference," I said and pointed to my suitcase.

"Good. Perfect. You can tell me if coffee causes cancer, and I'll give you free refills in exchange."

"That sign says you give everyone free refills."

"This is going to go a lot easier if you don't play hard to get. So does coffee cause cancer?"

"It's a bit complicated."

"Make it uncomplicated."

"It's kind of a long story."

"I'm not going anywhere." She rested her elbows on the countertop.

Truthfully, neither was I. "People have been talking about possible coffee-cancer links since the early 1980s."

"Wow, before I was born."

It's hard to meet people who think of the 1980s as pre-history. I felt like protesting that it was before I was born too, if only by a few months, but vanity has no place in medical research. "There was a study called 'Coffee and Cancer of the Pancreas.'"

"Pancreatic cancer is pretty bad, isn't it? Isn't that what Alex Trebek died of?"

It was. Pancreatic cancer is often aggressive and fatal, and rates of pancreatic cancer were going up throughout the 1970s. That's why Brian MacMahon and a group of researchers from the Harvard School of Public Health launched a research study and published "Coffee and Cancer of the Pancreas" in the *New England Journal of Medicine* in 1981. Given the rising cases, there was an urgent need to figure out what was causing the disease, so MacMahon and his research team designed a study where they could question people with pancreatic cancer about what they ate and drank. They hoped they could identify some patterns and explain why the cancer developed.

"So they, like what, experimented on them?" Casey asked.

"They had them fill out questionnaires. It was pretty non-invasive."

The study amounted to the researchers identifying pancreatic cancer patients in hospitals around the Boston and Rhode Island areas and having them fill out a food questionnaire. They got a control group to do the same and then compared the two to see if any specific food was more common in one group or the other. They thought the questionnaire would help identify, or at least suggest, if a specific food was causative or protective.

"So they just picked patients from the hospital they worked at?"

"Yep. I don't know if you've ever been in hospital . . ."

"I hate doctors. And hospitals. No offense."

I get that reaction a lot. "None taken."

"I do like looking at the newborn babies. But they don't let you do that anymore. Security said if I kept coming back, they'd call the police. Steve the security guard was really nice about it, but he said it was out of his hands. We're friends now. He sends me Christmas cards from his family. Beautiful kids."

I had a vision of Casey at the neonatal ward like Audrey Hepburn showing up to have breakfast while looking in at Tiffany's.

"Why are you smiling?" she asked.

"No reason." I changed the subject. "Despite your aversion to the non-baby parts of hospitals, they are good places to recruit patients for research studies."

"Glad they're useful for something besides my gynecologist talking down to me."

"That coffee study recruited patients between 1974 and 1979, in a world with no internet, no email, no social media, and no cell phones. It wasn't as easy to identify and contact patients."

"No social media. Wow. Imagine how much free time everybody had."

It was indeed a simpler time, and a lot of stuff took longer. There were no electronic databases to search, so the easiest way to find patients for your research was to go to your local hospital to find all the patients with the diagnosis you're looking for. Then you can use patients admitted to the hospital for other reasons as a control group.

What was interesting about MacMahon's study was that the initial hypothesis didn't pan out. They were initially looking for common risk factors like smoking or alcohol. In their analysis, smoking only had a minor effect while alcohol had none. Tea had none. But coffee had a very strong effect. People who drank one cup of coffee per day had twice the odds of getting pancreatic cancer. If you drank more than one cup per day, the odds tripled.

The study made a huge splash when it was published. The *New York Times* did a whole feature on it, and in an interview, MacMahon said coffee could be responsible for half of all pancreatic cancers.

Casey was searching on her phone. "I found it. 'Study Links Coffee Use to Pancreatic Cancer,' published in 1981. I can see how people would have lost their minds over this."

The results were striking. In their defense, the researchers did say we shouldn't over-interpret their findings and that a single study doesn't prove the link. But Dr. MacMahon also said he'd stopped drinking coffee because of the study. In the interview, he also made a link with Mormons and Seventh-Day Adventists. He attributed their lower rates of pancreatic cancer to their general avoidance of coffee.

"Why did people turn on him?" Casey was scrolling through her phone again. "There's another *New York Times* article here. 'Critics Say Coffee Study Was Flawed.'"

"Things unraveled pretty quickly."

"Why?"

"It's a concept called selection bias."

Casey said nothing.

"It means picking the wrong types of patients for your study."

"Like how women aren't included in studies and most research is only done on men?"

I tried to explain that what she'd just described was a different issue, but I was coming to realize that once Casey got started on a subject it was hard to stop her.

"Not that I really want to be experimented on or anything," she said. "In fact, I really don't. I don't need the money badly enough to take experimental drugs just to see if it's going to make my hair fall out. But it feels like a real jerk move to only do studies in men."

"You're not wrong."

"So why do it? Assuming there's a non-jerky reason to do it."

I thought about how I was going to explain this. "Imagine you want to see if smoking causes lung cancer."

"I feel like we know the answer to that one already."

"It's . . . we're just pretending."

"I can play make-believe with the best of them. Proceed."

"So —"

"For the purpose of this role play, my name is Cassandra. Just FYI."

There were many reasons why studies from decades back only included men. Some understandable. Some less so. It was for a time unfashionable for women to smoke, and few actually did so. But the driving motivation for many studies was the need to keep your exposure and control groups as similar as possible. If you want to compare smokers and non-smokers, then you want them to be identical in as many other ways as possible.

Casey listened patiently before asking the obvious question. "If you only study smoking in men, how would you know if it also causes cancer in women?"

"You wouldn't."

"Boo 19th-century science."

"This extended into the middle of the 20th century."

"Then boo that century too."

Her criticism, though succinct, was valid and widely shared. There is now a growing realization that you have to try to make your research groups as representative of the general population as possible. Casey seemed pleased the scientific community agreed with her, but she was immediately frustrated when I told her none of that had anything to do with selection bias.

"You sure like going on digressions, don't you?" she asked.

I did, but that was beside the point. "When we say *bias*, we mean it in the statistical sense, not in the way most people mean the word."

"Sounds unnecessarily confusing, but you do you."

"And we have two main types of bias: information bias and selection bias. Information bias is a problem with how you collect your data and selection bias is a problem with how you pick patients. It's funny, I was just discussing this."

"At the conference?"

"No, on the plane when I came here."

"With your doctor friends."

"No, with Jim and Katie. I met them on the plane. He works in insurance, and she does government work."

"You met them on the plane, and started talking to them about medical statistics?"

"Yes, well no. I met Jim before we took off. He was at the gate and needed some vitamin C, so I had to explain to him why that didn't work, and then we met Katie on the plane."

"And where are they now?"

"I think Jim is going to try and ask her out for dinner."

"Oh my God, is this a love story?" Casey shrieked. "That would be so amazing if they got married and had kids, and one day they told their grandkids they met because some weird doctor on a plane kept talking about vitamins and statistics, and that's how they fell in love." She sighed longingly.

"I think they hit it off, so I let them take a cab together so they could keep talking."

"About vitamins?"

"I don't know. Maybe. Probably not. I don't know."

She stared at me. "You're one weird dude." She gazed at me intently for a good long moment before making her final assessment. "I like it!"

I was pleased by her approval and told her so.

"Tell me about coffee and . . . that bias thing, which is not the same as racism in your world."

"Okay, let's say that we wanted to answer the question, 'Are taller people faster than shorter people?'"

"The answer to your question is, 'Who cares?'"

"I'm using it as an example. Bear with me. So imagine we wanted to study —"

"I'm going to go out on a limb here and say taller people are faster," Casey said. "They're taller, they have longer legs, they have a longer stride. It's obvious, isn't it?"

There was a funny story about that. When I was a resident, we used to have a stress test competition where we would see who could last the longest doing a standard stress test protocol. Most people can only last seven or eight minutes. A fit person can probably do 12. One year we had this medical student from France, and he wanted to try. He got on the treadmill, and he completely smoked us. He was wearing jeans and dress shoes, but he had these ridiculously long legs and could cover the whole treadmill in a single stride. At ten minutes he was still walking. He lasted longer than anybody I have ever seen. I told Casey the story.

"Must have been humiliating."

"I'm not going to lie. It stung."

"Did you ever win?"

"Yes, once." I still have the dollar store trophy on my bookshelf at work.

"Yay!" Casey was genuinely happy I'd won a pointless competition 10 years ago. She was that kind of person.

"The point of the story —"

"Not everything needs to have a point, you know."

In my world it does. "But this story does have a point. The point is taller people are, on average, faster than shorter people. Until you get to the NBA. Then the opposite is true."

"You're mistaken, good sir."

"Why are you speaking with a British accent?"

"I'm trying to tell you, not everything needs to have a point." She switched back to her normal voice so I didn't pursue the matter. "But you're wrong. Basketball players are both tall and fast. They prove the point."

"They don't. For one simple reason. Basketball players are different from the average person on the street."

"Agreed. They're richer."

They are richer. But that wasn't the point I was trying to make. This was much easier with Jim and Katie. The point was a subtle one about selection bias. Because in basketball players, the association between height and speed is the opposite of what you'd find in the general population. The weird thing about the NBA is the taller players tend to be slower. The shorter ones tend to be faster. Casey however wasn't buying it. She listened to my explanation patiently and intently. She considered my explanation, thought about it carefully, and responded with a single word.

"No."

I had to try a different tack. I sometimes use an analogy about the freshman class at a university and thought I'd give it a try. "There's an old joke in epidemiology about an epidemiologist who goes to visit his friend who teaches at the university."

"Is this the academic equivalent of 'two guys walk into a bar'?"

I didn't have a witty comeback, so I forged ahead. "So the epidemiologist goes to visit his friend who teaches at an Ivy League university."

"Which one?"

"It doesn't matter. Let's say Yale for the sake of argument. He gets there and asks one of the freshmen for directions. At that moment the epidemiologist's friend shows up, and the student says goodbye and leaves."

"This is a fascinating story so far, by the way."

"The epidemiologist says hello to his friend, and the friend asks him, 'Do you know who you were talking to?' And the epidemiologist answers, 'I assumed it was the son of a major donor.' 'How did you know that?' the friend asks. 'Well, there are only two ways to get into Yale. You either have to be very smart or have rich parents. He obviously had rich parents.'"

"Womp womp."

"Was that supposed to be a sad trombone?"

"It was."

"The point is —"

"You're hilarious. Everything has to have a point in your world, doesn't it?" She laughed.

Of course it did. The only other option was for things to be pointless. And what good was that to anyone? But I was on firmer ground with statistics than philosophy. And the story about that professor visiting a university revealed an important statistical concept. When you have two things that can lead to your desired outcome, if you can rule out one of them, the other must be true. I explained this to Casey.

She mulled it over as she drummed her fingers on the countertop. We had a slight reprieve when a customer came in and asked for a coffee to go. The pause had given Casey a chance to gather her thoughts and formulate a counterargument. "I'm not sure I believe you. I've known people who were both really smart *and* pretty well off, as well as people who were clearly neither."

In the general population that was probably true. A really smart person can invent something or start a new business and become super wealthy. In the wider world, you can easily be both or neither. But if you limit yourself just to first-year university students of an Ivy League school, everything skews the opposite way. The challenge was going to be explaining this to Casey in a way that made sense. I pulled a pen from my jacket pocket and drew a grid on a napkin.

"I didn't know there were going to be audiovisual aids," Casey said.

"Sometimes working with numbers helps."

"You and I have had very different life experiences if you're able to say that with a straight face."

I flipped the napkin around to show her what I'd written out.

	Rich	Not rich	Total
Total	50	50	100

I walked her through my example. "Let's assume we have a sample of 100 random people. If we picked them completely at random, 50 should be rich and the other 50 not." Technically, I should have said *above average* and *below average* wealth, but I was trying to keep things simple.

Casey had a different objection. "Income isn't actually evenly split in the population. Trust me. I have sooooooooo much student debt."

It was a fair point, but it didn't affect the example. The point still works mathematically even if the split is 70/30 or 80/20. The key is assuming that being rich and being smart are independent of each another. If you accept that assumption, then you can fill in more of the grid and you're left with four equal groups of people defined by how rich and how smart they are. If you assume everything is random, you have 25 people in each group.

	Rich	Not rich	Total
Smart	25	25	50
Not smart	25	25	50
Total	50	50	100

"With me so far?" I asked.

"Oh yeah. Totally."

Her smile might have been genuine, or she was laughing at me. I figured it was a coin toss, so I pressed ahead. "Now I'm going to change up the example."

"Is there going to be a dramatic and unexpected plot twist?"

I think she was messing with me, but she was pretty hard to read. "Take the same example, but now we limit ourselves to first-year students at Yale. Everything is the same, except one group gets left out."

"Somebody who's not smart or rich isn't getting into Yale. Because if they weren't rich enough to afford tuition, and not smart enough to earn a scholarship to pay for tuition, they won't get into an Ivy League school."

"Exactly!" I smacked my hand down on the counter and my coffee cup jumped and rattled on its saucer. It was admittedly an overreaction, but I get carried away with these concepts sometimes. I jumped at the clatter and Casey laughed at how easily I startled myself. I decided to ignore my outburst and turned the napkin around to show Casey the newest version of my little table.

	Rich	Not rich	Total
Smart	25	25	50
Not smart	25	X	25
Total	50	25	75

"So if you were speaking to an undergrad, and you realized he wasn't that smart, then there's only one possible explanation. He has to be rich."

"Or have rich parents," Casey said.

"Same thing in this world."

"So where do the basketball players come in?"

I took the napkin back, crossed out the headings and replaced *rich* with *fast* and *smart* with *tall*.

	~~Rich~~ Fast	~~Not rich~~ Slow	Total
~~Smart~~ Tall	25	25	50
~~Not smart~~ Short	25	X	25
Total	50	25	75

"It's the same basic idea," I said. "If you look at just NBA players, there's one group of people who will not be in your study sample."

"There are no short, slow players in the NBA."

"You got it."

Casey gave a fake sigh. "I should have been a doctor."

"Well . . ."

"I just never really liked the idea of getting someone else's blood on me."

"Sort of comes with the territory."

"I faint when I see blood. I'm like those fainting goats you see on YouTube."

I made a mental note to look that up later. I was going to keep going with the analogy, but Casey reached under the counter and brought up a large placemat that she put on the counter in front of me. "Just in case you want to draw more diagrams," she said.

I was pretty certain at this point, she was teasing me. "I think I made my point with this one."

"And what was the point exactly?"

"To make it in the NBA you have to be either very tall or very fast. I'd never make it in."

"Are you sure? You can do anything if you set your mind to it. At least that's what my high school guidance counselor used to say. He just never pointed out how important it was to be on the receiving end of nepotism."

I had no comeback. "You have to be tall and fast to make it into the NBA. That's a given. But if you see someone in the NBA who isn't that tall, then you can bet he must be pretty darn fast."

That was the point I was trying to make. In the general population, tall people are also fast. But if you made it in the NBA, a really tall person probably can get away with not being that fast and a really fast player can get away with not being that tall. In the NBA, the relationship between height and speed is reversed.

This was the core idea behind selection bias. By picking a particular subgroup of patients, you can introduce a bias into the association. You can make it seem as if taller people should be slow, smarter people should be poor, or that coffee causes cancer.

"I was wondering when you were going to get back to that," Casey said. "Refill? Since coffee doesn't actually cause cancer." She refilled my mug and reset the coffee pitcher on the burner. When she was done, she didn't say anything for a few seconds. After a long awkward pause she finally asked, "Well?"

"Well, what?"

"You didn't finish your explanation."

"I thought I had."

"Nope. You explained how in universities and the NBA selection bias made things look negative, like one thing repelled the other. But with coffee, it links two things together, as if coffee *causes* cancer."

There was a special type of selection bias called Berkson's bias. It comes into play when you study hospitalized patients, because hospitalized patients, just like NBA players and Ivy League university students, are different from the general population. Something has to happen to gain them entrance into that subset of the population. They either have to get drafted, get admitted, or get sick. I thought I might have some success drawing Casey out if I approached the issue tangentially.

"Let me ask you a question," I said.

"How many guesses do I get?"

"It's not that type of question. Why did people get enrolled in the study?"

"They had pancreatic cancer."

"But what about the control group? The non-pancreatic-cancer patients. They had to be in hospital to get recruited. So why were they there?"

"I dunno," she said. "Something other than pancreatic cancer. Heart disease, other cancers, broken bones, somebody shoved something up someone else's butt? There's a whole show about that on TV. It's awesome."

I don't watch much reality TV, but I found it hard to believe that anyone would want to watch something like that. Casey assured me it was one of the more popular shows on TV.

"You have to realize that hospitals tend to put patients together on wards," I said. "The cardiac patients are together on one floor, the cancer patients on another and so on." Casey agreed it was a logical setup. "So when MacMahon and his team recruited the pancreatic cancer patients, the other patients on the ward got recruited as controls."

"I'm still waiting for the punchline," she said.

"The punchline is they had stomach ulcers."

"People don't get hospitalized because they have stomach ulcers."

"Not now but back then they did. In the '70s and '80s the treatment was often a partial gastrectomy."

"And what, pray tell, is a partial gastrectomy?" She'd become British again.

"It's when you cut away a part of the stomach."

"What?!" She wasn't British anymore. "Are you kidding me?!? They would carve out part of your stomach . . . because of an ulcer?!? What kind of crazy world were we living in!"

I thought back to the Cold War, apartheid, and the AIDS epidemic. "Can't argue with you there. There were a lot of low points in those decades."

Today's patients will rarely need to have surgery for a stomach ulcer. Medical advances have made that obsolete. But at the time MacMahon was doing his coffee study, such treatments were not available, and our basic understanding of what caused stomach ulcers was fundamentally wrong. Ask anyone from that time and they would likely tell you that stomach ulcers are caused by stress. Many people today still tell you that when asked. But thanks in large part to the work of Barry Marshall, we now know that most stomach ulcers are caused by a bacterial infection.

In the same year that MacMahon published his paper linking coffee and pancreatic cancer, halfway across the world in Australia, an internal medicine resident named Barry Marshall was handed a list of patients by Robin Warren, a pathologist at Royal Perth Hospital. Warren had discovered a corkscrew-shaped bacterium in the stomach biopsies of some hospital patients, and Marshall was given the task of figuring out what, if anything, this meant. Things didn't initially go well. Marshall tried to infect pigs, mice, and rats with *H. pylori* to see if they developed stomach ulcers. They didn't.

"Bit of a setback," Casey said.

I agreed. "Marshall realized if he was going to prove his theory, he would have to use humans, not animals."

"Humans are animals, but I get your point. So where did he find research subjects?"

"He had to start small, and he experimented on the first human he could find."

"You mean . . . ?"

"He experimented on himself."

"Brave. Kind of stupid. But brave."

"This has become medical folklore. Marshall drank a broth of *H. pylori* bacteria to see if it would cause stomach ulcers. Ten days into the experiment, he was throwing up every morning. His wife found him in the bathroom one morning and got the story out of him. She forced him to take antibiotics and reverse the process."

"She sounds like a sensible woman. I think we would be friends."

That was the start. Marshall had proved the hypothesis. Infecting himself caused the problem. Killing the bacteria with antibiotics cured him. It's hard to overstate how important this discovery was. It radically changed how doctors approached not just stomach ulcers but many other related problems like gastritis and even stomach cancer. Not everyone with a stomach ulcer has *H. pylori*. They can be caused by other things like anti-inflammatories or steroids. But the realization that many stomach ulcers were caused by bacteria and not by stress or spicy foods changed the practice of medicine. Marshall and Warren won the Nobel Prize for their discovery, and countless patients were spared unnecessary surgery.

Casey said that she liked the story because it had a happy ending. But she had one quibble. "My uncle who had ulcers often said he felt worse if he had very spicy food or drank a lot of alcohol."

"I'm skeptical. Most foods don't make a difference because no matter how spicy or acidic your food is, it's not going to be more acidic than the acid you make in your stomach."

"So my uncle was just full of it?"

"Well, I don't know your uncle."

"That's for the best. He was kind of racist."

Casey's uncle was probably one of the many people who grew up hearing ulcers were caused by stress. And like most people he was probably told to avoid things like alcohol and cigarettes in an effort to relieve the stress on his stomach. That advice might not have been entirely misguided. Smoking does seem to increase the risk of peptic ulcer disease, though the data for alcohol is less clear. But there's no evidence any specific food, spicy or otherwise, has any impact. And yet it's still very common for people with ulcer and gastrointestinal diseases in general to avoid specific foods. Especially coffee.

"Yup, that's me," Casey said. "Coffee makes my stomach all fluttery."

"Coffee may not cause stomach ulcers, but it can cause dyspepsia. An upset stomach."

"I see where you're going with this."

"Okay, tell me." I sat back and ceded the floor. I have to admit, I was impressed. Casey got it mostly right. She'd seen the point I was driving at and was able to grasp the significance. If people with stomach issues drink less coffee, you're going to have a problem if you include them in your control group. MacMahon's research group recruited their control patients from the same hospital wards where they found the pancreatic cancer patients. So many of those patients had their own underlying issues like stomach ulcers, gastritis, colitis, or irritable bowel syndrome. So it wasn't that pancreatic cancer patients drank more coffee. The control group drank less.

I told Casey she was basically right, and she took great pride in the fact that she was smarter than a bunch of Harvard scientists. But the triumph was short-lived because she started to look a little pale and unsteady on her feet.

"I need to go sit down," she said.

Worried she might faint, I got up to steady her. I walked her over to the couch and told her to lie down. "Better elevate your legs."

"Boy, if I had a nickel for every time I heard that from a guy I just met."

I can't imagine what my face looked like because Casey started laughing.

"I'll be okay in a few minutes. I get like this sometimes. We were just talking about people having cancer and colitis and IBS, and I don't like talking about people getting sick or being in the hospital. I told you, I'm like those goats. Anything upsetting or stressful and I just go down."

"It's okay. Let me get you some water." I filled a glass from the tap and on my way back, I flipped the sign on the door. It wasn't locked but with CLOSED facing out, I figured that would dissuade most people.

"I'll just lie here for a bit. But if my manager comes, you were the one who fainted and *I* helped *you*."

I assured her that she could count on me.

"Try to use the word *heroic* when you describe my actions."

"Your color's starting to come back," I said.

"Just keep talking to me. I like the sound of your voice. I don't understand half of what you're saying, but your voice is soothing. You're like a white noise machine that's trying to teach me math."

I thought she was keeping up the conversation quite well, so I wasn't sure how much of that last part was true.

"I don't remember what we were saying," I said.

"It wasn't that the cancer patients drank more coffee. The control group just drank less because they had other stuff going on."

It was as succinct a summary as I could have provided. I went back to the counter and picked up the napkin I'd sketched things on before. I crossed out my previous edits and renamed everything once again. It was the same table except now *Tall* was *Coffee* and *Fast* was *Cancer*. The concept was the same. Only the names had changed. And the critical part was that one of the four groups was left out of the analysis. The non-cancer patients who drank regular amounts of coffee were absent because they weren't hospitalized when the researchers went recruiting.

	~~Rich Fast~~ Pancreatic Cancer	~~Not rich Slow~~ Control Group	Total
~~Smart Tall~~ Coffee	25	X	25
~~Not smart Short~~ No Coffee	25	25	50
Total	50	25	75

Casey looked over the napkin and proclaimed her approval. She sat up, sipped some of the water I'd brought her, and looked over the napkin again. "You know, when you say it, it seems obvious." I shrugged. "I can't understand how people make mistakes like this."

"People make this mistake every day," I said.

She told me she needed an example as she tried to get up. She made it but then wobbled a little bit and sat back down on the couch. "Nope. We're doing this from down here."

"Can I get you anything?"

"I hate having low blood pressure. Maybe I need to eat more salt."

I restrained myself from commenting. One thing at a time. "We make the mistake of selection bias every day without realizing it."

"How?"

"For example, do you do online dating?"

"Sure. Tinder."

"And what motivates you to make a decision on someone?"

"Desperation and the fear of dying alone."

I choked a little on the water I was drinking. She was either deathly serious or the best deadpan comedian I had seen.

"There aren't many options on the apps these days and there's a lot of low cards in that deck," Casey said. "You're lucky you're married and don't have to go on dates anymore."

As I got my coughing fit under control, I told her I wasn't married. I also realized I was going to have to be more specific in my analogy. I tried a different approach. "I bet you think there are nice guys and good-looking guys, but you can't find someone who's both."

"I can tell you from personal experience that is absolutely true. Good-looking guys are jerks and vice versa."

"But consider this. You probably haven't dated every single guy in the city."

"That feels like an insult."

This was going poorly. I tried again and walked Casey through another theoretical scenario. I asked her to pretend she was in a situation where she had 100 eligible bachelors to choose from. Casey said that sounded like a great reality show she would definitely watch. I ignored the comment.

The problem was that the 100 imaginary bachelors weren't a random cross section of society. If you agree to go out on a date with someone, there has to be a reason. Maybe that person has a great personality. Maybe he's a good listener, or funny, or charming, or a kind-hearted soul. Or maybe he's tall, striking, with smoldering good looks, a chiseled jaw, and washboard abs. Or maybe he's filthy rich with a huge house, fancy cars, and money to burn.

Casey interrupted me to say she voted for option B. She then changed her vote to C, then thought better of it and went back to B again. I told her there was no voting, and she pouted a little. The point of my little analogy was you go out with someone because they check off some box. They couldn't fall short in all respects otherwise you wouldn't go out with them. Casey was still dubious.

"Would you really go out with someone who is both ugly and a jerk?" I asked.

"I really want to answer 'no' to that question. I really do."

I'm hardly someone who should be handing out relationship advice, but dating has the same problem as university admissions and being hospitalized. There has to be a reason you made it into that select group of people. Hospitalized patients are sick with something,

Ivy League university undergrads either paid the tuition or earned a scholarship, and someone you agreed to go out with had something working in their favor. They were either kind or good-looking. They were possibly both, but they probably weren't neither. Selection bias was at play here too. Since the unattractive people with bad personalities are underrepresented in your dating pool, you make the flawed assumption that looks and personality are inversely related. Casey seemed to accept what I was saying, but she also had a counterargument based on what I assumed was real-world experience.

"It could also be that the good-looking and kind ones were snatched up early and are not on the market anymore," she said.

"That's possibly true. A bit of cliché, but I guess true to some extent."

She cocked her head to one side and looked at me intently. "How come you never got married?" she asked.

"I work 100 hours a week on average. Just never found time for it."

She nodded. "Ah, one of those."

"But even if we accept the premise of your suggestion, the point still holds."

I reached for my pen to make some more sketches on the napkin, but I realized it fell on the floor when I helped Casey to the couch.

Casey followed my gaze. "Are there more graphs?"

I told her there were.

"Then help me up. It's okay, I'm feeling better."

I helped her to her feet, and we sat on stools by the counter. I found my napkin and sketched out another two-by-two table.

	Attractive	Unattractive
Kind	X	✓
Not Kind	✓	X

"Even if we remove the ones who are kind and attractive, the problem remains. Looks and personality look like they're opposites because you've excluded the people where they track together."

"That doesn't make dating any easier," Casey said.

It didn't but that was outside my purview. Casey studied the napkin and twirled it around the countertop. "You make it sound like this is a common problem."

"It is a bit."

"So why don't people spot it? How come nobody said anything when that coffee study was published?"

A lot of people did. After a while, old manuscripts move out from behind their paywall. I called them up on my phone and showed Casey the flurry of correspondence that grew out of the original paper. A letter from Dr. Steven Shedlofsky asked, "I'd like to know how many of the 32 control patients with no coffee intake had any of the above gastrointestinal problems and whether they avoided coffee because of the illness." He'd realized almost immediately how selection bias could skew the results.

About a month after the paper was published, Dr. Harvey Goldstein from the Scripps Clinic in California sent a letter to the journal where he described the results of a survey he'd done with his patients. He took 91 patients with pancreatic cancer and 93 control patients that had either breast or prostate cancer. The important difference here was his control group had other forms of cancer, not gastrointestinal problems that might have affected their coffee consumption. He found no link between coffee and pancreatic cancer. There were the same number of coffee drinkers in both groups.

Later that year there was an editorial in the *Journal of the American Medical Association* that also pointed out the problem of how the controls were chosen. They said the paper and the press coverage it got were like yelling "Fire!" in a crowded theater.

"Harsh," Casey said as she scrolled through the articles. "Couldn't they just go back and fix the paper?"

To an outsider, fixing a scientific paper sounds easy. In practice, it's quite hard. MacMahon and his group did address some of these problems in a limited way. Whenever someone writes a letter to the editor about your study, you have to respond to it, and in the replies, MacMahon acknowledged the problem of the control patients having gastrointestinal problems. He also gave an interview to the *Epi Monitor* newsletter where he conceded the point without backing down. They also did some further analyses on their data where they excluded the control patients with gastrointestinal problems. In that analysis, coffee wasn't dangerous.

Similar papers were published in *Cancer Research* in 1983 and *Cancer* in 1985. Neither found a link between coffee and pancreatic cancer. But they had a different strategy for picking their control group. They took their patients with pancreatic cancer and scrambled the last four digits of their landlines. They then called those numbers and invited them to participate. The method not only ensured that the controls were picked essentially at random, but it also had the advantage of picking people from the same neighborhood as the cases because the first three digits of their phone numbers were the same. Say what you will about landlines, and Casey had plenty to say about them, they did make research a little bit easier.

Following a 10-minute soliloquy from Casey about why cell phones were better than landlines in every conceivable way, I tried to bring up the cost but was shushed. Casey's rant continued until an idea occurred to her, seemingly mid-sentence. "Did MacMahon and his group ever do a follow-up study?"

"They did. In 1986. It was called 'Coffee and Pancreatic Cancer (Chapter 2).'"

"Catchy title. Bit on the nose. Did it show anything?"

"No. And by that point, the evidence was piling up, so they just admitted their initial study was probably wrong."

After this second study was published, the *LA Times* ran a story with the headline, "Drink Up: New Research Disputes Reported Link Between Coffee, Cancer."

"I like happy endings," Casey said. "So are you a dog person or a cat person?"

"Dog person, I suppose. But I don't have —"

"Wait a minute. What about the labels?"

"Labels?"

"We started talking about all this because you said coffee had to have warning labels. Or maybe I said it. I can't remember."

I'd genuinely forgotten what prompted this whole conversation — the case and judgment requiring coffee to have a warning label. "That's all because of Proposition 65," I said.

Casey nodded. "It's supposed to keep toxic stuff out of our food, isn't it?"

"Sort of."

Prop 65 is officially called the Safe Drinking Water and Toxic Enforcement Act, and it requires the state of California to keep an up-to-date list of things that could cause cancer or birth defects. On the surface, it's hard to argue against an initiative like that. The list includes things like asbestos, tobacco, and the smoke produced from burning coal. Casey pointed out that these things are well-known to be bad for you, and we don't need a government department to tell us that. But at the other end of the spectrum are things whose danger to humans is very questionable. Medications like levodopa are also on the list. Casey's grandfather had Parkinson's disease, so she knew what the medication was for.

"It's kind of weird that they would include it. Obviously, medications have side effects, but it's not like people are putting this in the drinking water. It's being prescribed to sick people."

"The reason it's on the list is because it can cross the placenta so it could cause birth defects if a woman took it while pregnant."

"Kaaaaaaaaay . . . Except most women aren't getting pregnant in their 70s. I feel like there's not much overlap between people who get pregnant and people who get Parkinson's disease. This may not be the greatest public health threat our species has ever faced."

I conceded that she was right.

"What else is on the list?" she asked.

"Alcohol."

"I think we tried to make alcohol illegal at some point. It didn't go well. The Ken Burns documentary was awesome."

"Prohibition did turn out very badly." I'd also seen the Ken Burns documentary and agreed that it was excellent.

"I feel like Proposition 65 is sort of undercutting its authority if it includes common medications and alcohol on its list."

"You know what else is on the list? Estrogen." Casey cocked her head. "Because it causes breast cancer," I explained.

"Did anyone stop and think that every woman has estrogen inside them and that labeling estrogen as a carcinogen seems a little silly?"

"Actually, men have estrogen inside them too, to be accurate. But I take your point."

"So why is coffee on the Proposition 65 list?"

"It isn't. But acrylamide is."

"And what's that?"

"It's a chemical that's used to make paper, dyes, and plastics. Stuff like that."

"And it causes cancer?"

"In rats, yes."

"But I'm not a rat," Casey quite reasonably pointed out.

"In humans the risk is less certain. In high doses, potentially. But in the doses that most people get in their diet, it's unlikely."

"I'm a little bit worried that a possible cancer-causing chemical is in the food I eat."

"That's understandable. But acrylamide occurs naturally in food, and the doses are so small it's not likely to be hurting anyone."

I wouldn't blame anyone for worrying that they were eating something carcinogenic. But acrylamide isn't some toxic substance being added to our food. It's naturally present in many foods at very low and essentially harmless levels. Heating or cooking food converts the amino acid asparagine into acrylamide, and coffee isn't immune from the process. Roasting coffee beans produces acrylamide and has done so for as long as humans have been drinking the beverage.

The California lawsuit was brought forward by the Council for Education and Research on Toxics, a private non-profit group that has sued many retailers in the past over Proposition 65 cases. They sued over 90 coffee retailers like Dunkin' Donuts and Starbucks alleging that since coffee contains acrylamide, selling it without a warning label violated Proposition 65.

Several companies, like the 7-Eleven chain, settled out of court and avoided the costs of a trial. Starbucks chose to fight the lawsuit. They ultimately lost after Superior Court Judge Elihu Berle ruled, "Defendants failed to satisfy their burden of proving by a preponderance of evidence that consumption of coffee confers a benefit to human health." But in a surprise twist, in June of 2018, California's Office of Environmental Health and Hazard Assessment (OEHHA) ruled that no warning label on coffee would be required. After reviewing the scientific evidence, they stated, "Exposures to listed chemicals in coffee created by roasting coffee beans or brewing coffee do not pose a significant risk of cancer." The OEHHA ruling changed everything. Now armed with fresh ammunition, Starbucks and its co-defendants asked for and received a summary judgment from Judge Berle in 2020. In short, they won.

"I wondered why we never had to put warning labels on the cups," Casey said.

"Sometimes things work out the way they should."

"My life experience doesn't suggest that's true. But I think your optimism is cute and charming."

"I still like to believe that when the chips are down people will, usually, do the right thing. And that's what the OEHHA did. They reviewed all the evidence and looked at the 2016 WHO report and came to the logical conclusion."

Casey nodded along but then stopped. "Wait a minute. Didn't the World Health Organization report say the opposite? That coffee was cancerous? That was the *Time* magazine article we were talking about. Remember?" She called up the article on her phone again and showed it to me as proof.

"That headline is a bit misleading. First, the WHO actually downgraded coffee in that report."

"What do you mean downgraded?"

"So the World Health Organization has an agency called the International Agency for Research on Cancer, or IARC, and their job is to classify stuff as either carcinogenic or not. So they can classify products as Group 1 (carcinogenic to humans), Group 2A (probably carcinogenic), Group 2B (possibly carcinogenic), Group 3 (unclassifiable), or Group 4 (probably not carcinogenic)."

"Was this what you were discussing with your friends on the plane?" she asked with half a smile.

"It was, actually."

"Is that how you make people fall in love with each other?" She rested her elbows on the counter, leaned forward and rested her head on her hand in a theatrical pose while she fluttered her eyelashes. She giggled.

I wasn't sure if she was laughing at my expense.

"So what happened with coffee?" she asked.

"Well in their previous review in 1991, IARC put coffee in Group 2B, meaning possibly carcinogenic."

"That seems a bit wishy-washy. Anything is 'possible.' By that logic you could put almost anything in Group 2B."

"Trust me you're not the first person to point that out. But in the most recent report, they moved coffee from Group 2B to Group 3. They downgraded it because the past 20 years had produced a lot of research showing no coffee-cancer link."

Casey pursed her lips in a confused pout. "So . . . why the *Time* article then?"

"There was a second part to the IARC report. The cancer link had nothing to do with coffee but with maté."

"Tried that once. My friend Sofia introduced me to it."

"Do you sell it here?"

"No," replied Casey a little forlornly. "I brought it up once, but management wasn't keen. I'm not sure what their issue was. I mean it's a hot drink with caffeine in it. I feel like that's what we're all about."

"I guess it's less popular here than it is in South America."

"It's pretty popular here too. Don't worry."

I wasn't very worried, but I appreciated her reassurance.

"But what does maté have to do with cancer?" Casey asked.

"The IARC review found two studies linking hot maté with cancer of the esophagus."

Casey subconsciously raised her hand to her throat. "So throat cancer?"

I nodded.

"Are we about to go through a whole thing where the problems with the coffee study happened again with these two maté studies?"

It's always a bit disappointing when somebody steals your punchline, but that is essentially what happened. Researchers identified patients with esophageal cancer and questioned them about their eating habits and compared them against controls. Since the esophageal cancer patients were all admitted to local hospitals, when it came time to pick the controls, they did the most natural thing in the world. They picked their controls from among the patients hospitalized for other reasons. Just like MacMahon did with his coffee study.

"The worst part is that it's not even about maté," I went on. "When IARC looked at the published trials, of which there were frankly not that many, the association was only seen with hot maté, not with maté served cold."

"So not the drink itself?"

"They specifically pointed the finger at maté that was served hot and then sort of extrapolated to all hot beverages. Maté served cold or 'not very hot,' as they put it, was classified as Group 3 or 'not classifiable' as a carcinogen."

"And maté served very hot was listed as what?"

"It and all hot beverages were listed as Group 2A. Probably carcinogenic to humans."

Casey contemplated this information and looked long and hard at the pot of coffee she was keeping on the burner behind the counter. I could only guess at what was passing through her head.

"I mean, I guess drinking scalding liquid could damage your throat," Casey said slowly, but there was a hint of skepticism in her voice. "I guess that probably makes sense. But those headlines really seemed to be talking about coffee, not hot liquids in general. By that logic, it should apply to tea as well, shouldn't it?"

"It should."

"So it's not coffee, but hot drinks." Casey seemed a bit mollified. "Which begs the question, how hot is hot?"

"In the IARC report, the Group 2A label was given to hot beverages that were drunk at or above 65 degrees Celsius."

"And what is that on a real thermometer?"

"You know most of the world uses the metric system now."

"Most of the world isn't awesome. So in Fahrenheit that would be?"

I took out my phone. "149 degrees Fahrenheit."

"That seems really hot. I don't think we serve it that hot. That would burn a person."

"We could test it."

"You want me to try and scald you with hot coffee? You're into some kinky stuff."

"No! Good God! No. I mean make a new cup of coffee, and we'll measure the temperature of it."

"I don't have a thermometer. Rarely comes up at work."

"I have one in my bag. It's part of the medical kit I always have with me when I travel. You make a new coffee and I'll see how hot it is."

Casey went behind the counter to brew another pot of coffee and I fished the small black pouch out of my bag. When the coffee was ready, I put the thermometer into the steaming mug and we both stared at it.

"This isn't how I thought I was going to be spending my day," Casey said.

We sat there in silence while the thermometer did its thing. Strangely, neither of us felt the need to say anything. After a few moments, the thermometer beeped. "It says 52 degrees Celsius."

Casey cleared her throat. Fortunately, the thermometer reported both. "Which would be 125 Fahrenheit."

"And given that most people sip their coffee slowly, it's probably getting colder by the second as it sits on the counter," she said.

"It must."

"So, you don't think there's anything to the idea that coffee can increase your risk of cancer?"

I reached forward, took the cup, and sipped it slowly. As I swallowed, I tapped the thermometer on the countertop. "Nope, not hot enough. Under these conditions, IARC wouldn't be worried about this cup of coffee causing cancer. And quite honestly, neither am I."

"That's good," she said. She leaned her elbows on the counter again and rested her head in her hands. "Your dying would ruin my day. You're funny." She giggled.

I didn't think I said anything particularly funny, but I assumed it was a compliment, so I took it as such. I sipped my coffee, and we talked about other things. Casey had many follow-up questions about Jim and Katie. At one point I got a text message from the hotel saying that my room was ready. I silenced the notification and put the phone back in my pocket. I was in no hurry.

We also completely forgot to flip the door sign back.

MYTH #5

Red Wine's Good for Your Heart

I stayed in the coffee shop longer than I should have, and it was later than I'd planned when I finally checked into my hotel room. By the time I unpacked my bag, I was bone tired and crawled into bed for what I thought was going to be a quick 15-minute power nap.

About an hour later, my rather inexplicable dream about a talking dog named Finnegan was interrupted. My phone buzzed on the dresser by my bed. I fumbled for the offending noise maker. My friend Alexi had messaged me.

Alexi and I had done our medical training together, but life and circumstance conspired to separate us as we took on different specialties. He'd gone off to Italy to do some additional training in Milan, where he developed a taste for Italian suits and Italian wine. He also came back with an Italian fiancée to complete the package.

Since we were both going to be in town for the conference, we'd agreed to meet for dinner. He texted to see if we were still on. Under other circumstances, I would have been tempted to beg off. I was drained from the trip, but opportunities to reconnect with old friends were getting fewer and further between. So I hauled myself out of bed, showered, shaved, and headed down to the restaurant.

We embraced warmly and agreed it had been too long.

"You haven't changed at all!" he said.

I wasn't entirely sure that was true. Looking in the mirror after the shower, I'd seen a single gray hair weaving its way among its darker brethren. But I suppose I wasn't too far gone just yet. At least not in Alexi's eyes. He hadn't changed much either and I told him so.

"Yes, I have," he said. "Work has aged me ten years!"

Given that we hadn't seen each other in five years, that didn't seem too bad, all things considered.

"You're still a heartbreaker in my eyes, don't worry," I assured him.

He swore and informed me that I was still a jackass in *his* eyes. "Are you ready for tomorrow?"

I was giving a presentation at one of the conference sessions. "Yup."

"All done?" He was surprised.

Most people tend to procrastinate and work on their presentations the night before their talks in a flurry of stress and anxiety. "Yes, it's been ready for a while."

"Of course it has." He laughed. "I'd expect nothing less."

The waiter arrived to take our orders, and Alexi took the liberty of ordering the wine. He felt strongly that we should go with a particular selection. His arguments went over my head. I told him as much.

"Do you basically just choose between red and white wine?" he asked.

"That's my usual dividing line."

"I'll show you how to appreciate the finer things in life."

"You know in blind taste tests, most people show equal preference for a $20 bottle of wine and a $200 bottle of wine."

Alexi looked offended. "Yes, well," he said, "most people are wrong."

Our wine arrived and the waiter opened the bottle for us. He offered me the cork but since sniffing it would have been pure pretense on my part, I waved it off, and he offered it to Alexi, who sniffed it and pronounced himself satisfied. The waiter then poured a small sample into the glass and Alexi sipped it, again pronouncing himself satisfied with the result.

"Out of curiosity," I asked the waiter, "does anyone ever send the wine back after they taste it?"

He thought about it as he poured. "Every so often, yes. It maybe happens to me once a year and it's probably the same for everyone else here too."

"Because they don't like the taste?" I asked.

The waiter shook his head. "Almost always because the wine has gone bad. We store it downstairs in a cooled room, obviously, but every so often a bottle will spoil. It happens. It's the cost of doing business. We just get rid of it and get another."

"There's no such thing as bad wine," Alexi told me as he raised his glass, "only vinegar trying to pass itself off as wine." We toasted each other's good health. "Now that my return is imminent, we'll have to make it a point to see each other more often. It'll be just like the old days. I'll give you the benefit of my oenological experience."

Maybe Heraclitus was wrong. Maybe you can step in the same river twice. It would be nice if that were true. "Might be a wasted effort on your part."

"You have to drink the right wine," he said. "You didn't drink that swill they serve on airplanes, did you?"

I laughed and told him I didn't.

"Good, because you can ruin your palate with stuff like that. Although I suppose it's better than nothing, and it helps pass the time during the flight."

I told Alexi about Jim and Katie and our conversation about meat and salt and the finer points of epidemiology. He listened as he sipped his wine. "Was she pretty?"

I knew that would be his first question. I smiled despite myself and told him that she was in fact quite pretty.

"You should have asked her out."

"I'm not looking to date anyone right now. Plus Jim beat me to the punch. At least I think he did. I don't know how it turned out."

"You should ask him," Alexi told me.

"Ask him?"

"Text him and ask how it went. We need to find out how this story ends."

I had my doubts about this plan, but part of me was a little bit curious, so I fired off a quick message to the number Jim had given me. Alexi topped up our glasses as I typed.

"I may have to go easy on that," I said. "It was a long flight and a long day. I was sitting in a coffee shop all day."

"Pretty barista?"

She was, but I wasn't going to give him the satisfaction of telling him so. When I waved off his next attempt to top up my glass, he looked quizzical.

"I'm trying to limit my alcohol intake."

"Not buying into the heart-healthy aspects of red wine?"

"Wine isn't good for your heart."

"A lot of people disagree with you."

"A lot of people are wrong," I told him, echoing his own comment back to him. I couldn't tell if he disagreed with me or was just needling me for his own amusement.

"Do you get a lot of your patients asking you about red wine?" he asked.

"Quite a few. Check out this cartoon I use for my lecture on the subject." I showed him on my phone.

CARTOONIST: LUC SPERL

"This actually happens?" he asked.

"Often enough."

"People force themselves to drink red wine even though they don't want to?"

"You'd be surprised."

Wine being good for your heart is one of those things that everybody has heard about but nobody remembers from where. There's no shortage of research regarding alcohol's effect on the heart, but the widespread public buy-in on the issue came from an unlikely source. A *60 Minutes* story that aired back in 1991.

"So everybody watched a TV show and started chugging red wine?"

He laughed, but it was true. After the show aired, sales of red wine skyrocketed. People started talking about wine as if it was health food.

Alexi waved over the waiter as we were talking and took the liberty of ordering a cheese platter as an appetizer. He said you have to pair the right cheese with the right wine. I smiled and said nothing because before he went overseas the right cheese came in a can and the right wine was any wine.

"But the *60 Minutes* story must have been based on something," he said.

"The segment centered on the research of Serge Renaud, who was investigating heart disease rates in France. But Morley Safer had the most memorable line. He said the answer to the riddle of the French paradox may lie in the inviting glass of red wine sitting before him."

"How very poetic," Alexi said as he swirled his glass of wine. "By the way, you're not supposed to hold the wine glass like that."

I was holding my glass fully in my palm.

"You hold it by the stem so that your hand doesn't warm up the wine. That's why wine glasses are made like that."

"Is that true?"

"It's true tonight." He took another sip.

He seemed to feel strongly about the issue, so I put the glass down and took it by the stem. "So that's where that comes from," I said. "One interview with Morley Safer on *60 Minutes*, and we're off to the races."

The interview itself is a bit dated. The segment introduced TV viewers to the French paradox, the head-scratching observation that the French had less heart disease than Americans or Brits despite eating a high-fat diet. Everyone remembers red wine as the explanation, but the segment itself suggested a few other culprits. Renaud spent as much time blaming milk. Children in France drank very little milk relative to their American counterparts. And it was never entirely clear why alcohol should be good for your heart. People today invoke the notion that alcohol raises your good cholesterol. But that argument came later. At the time, when Renaud published his paper in the *Lancet* in 1992, he was looking at whether alcohol could inhibit platelets and thin the blood.

"We see that post-op," Alexi said. "The patients who drink a lot bleed more."

"What's the cutoff you use?"

"In terms of the number of drinks?" He sat back and looked at the ceiling. "I don't think we have a firm cutoff, but patients who drink *a lot* do worse afterward. Everyone sees that."

I nodded at the bottle. "All the more reason to cut back."

"I'm not planning to operate on you tonight." Alexi took a sip and swished it around his mouth before swallowing. "Why do you push back against the French paradox thing so much?"

We all have our little obsessions and I guess that was mine. My objection is the 30 years' worth of research that goes against the hypothesis. But even contemporaries had objections. Many researchers at the time argued that the lower rates of heart disease in France weren't due to wine but because the French ate more vegetables. Alexi accepted my arguments silently, but his main takeaway when I mentioned vegetables was that we should probably order.

"What looks good to you?" I asked.

"Well, if we want to get something that will pair nicely with the wine —"

"I think you make some of these things up."

"I do sometimes."

We spent the next 15 minutes debating what to eat since we'd tacitly decided that a mutual veto was now in effect. After nixing pasta and ruling out fish because of the wine we were drinking, we settled on veal. I just had to remember not to mention this to Katie should she or Jim ever get in touch.

"That was exhausting," I told him. "People don't usually spend this much time ordering."

Alexi refilled our glasses. "Wine will help. Maybe it explains the French paradox. Maybe not. The answer might be at the bottom of this bottle."

"There's a lot of differences between the French and North American diets other than wine."

"True. Serving sizes are much smaller in Europe. It shocks me when I come back. I forget how big plates are over here."

I started counting off differences between France and America on my fingers. "Smaller portion sizes, more fruits and vegetables, less junk food, less sugar, people walk more, and they have lower rates of obesity. The French paradox isn't that paradoxical."

I left a lot out. Then there was the debate about whether the differences between France and the U.S. had to do with how medical statistics are recorded and calculated in the two countries. If someone suffered a fatal arrhythmia because of a heart attack, what would someone record as the cause of death? How those deaths are recorded differs between the U.S. and France and could explain at least some of the variability. Not all of it, but possibly 10 to 20 percent. Alexi demurred to debate the point. It had been too long since he'd filled out a death certificate. He claimed it was because he was such a good surgeon that none of his patients died. I suspected he had underlings do it for him.

Another possible explanation for the French paradox was the time-lag hypothesis. Its basic premise argues that heart disease takes decades to develop. So if you wanted to look at heart disease rates in the 1980s, you should probably be looking at dietary patterns from the 1950s, not the dietary patterns of the 1980s. When you do that, France is much less of an outlier compared to other European countries and the U.S. Alexi had heard all these arguments from me before, but he rarely engaged or abandoned his position of aloof neutrality. "I know alcohol isn't good for me," he'd told me once, "but living in semi-denial serves me well."

When our food arrived, our conversation turned to our appreciation of the chef's skills and a similar dinner we'd had in Chicago when we attended our first medical conference together almost a decade ago. We had lit up the town.

"Let's get another bottle," Alexi said as he emptied out the last of the first one.

"I don't know."

"Are you driving anywhere?"

"No."

"You pregnant?"

"Not the last time I checked."

"Alright then." He ordered a bottle from the Bordeaux region. "To honor Renaud and the French paradox."

"He passed away several years ago," I pointed out. "I think he's beyond caring what we think of him."

"Then we'll drink to long-lost friends and the people we don't see anymore."

When the new bottle arrived, we proceeded to do exactly that.

Alexi vetted the new bottle. "There's still one thing I don't understand about all this."

"You use the corkscrew as a lever to pull the cork out of the bottle's neck," I said.

He snorted. "I mean about the French paradox."

"What's that?"

"Even if you discount the initial paper and the *60 Minutes* report as being a bit simplistic, there's still a lot of research about wine being cardioprotective."

"Sure, if you believe those papers."

"What made you so cynical?"

"Spending so much time with you, obviously."

He grumbled something about my mother saying he was a bad influence, which, of course, he was. He asked about my parents, who were still in prime health and close to retirement. I commented that his must be happy to have him back soon.

"They are. They weren't expecting me to come back with a girlfriend."

"Fiancée," I corrected him.

"Right, I have to stop saying *girlfriend*."

"Ideally, get into the habit before the wedding." I was taking a sip when a thought occurred to me. "Which is when by the way?"

"May. Before the heat."

"Smart. You don't want to be sweating in your suit."

"Exactly." This is how men think about wedding planning. We had nothing else to say about the wedding, so Alexi swirled his wine and looked at the little eddy he was forming in the center of the glass. "So is this whole red wine French paradox thing even relevant anymore?"

"How do you mean?"

"Well, the whole point originally was to figure out why the heart disease rates were lower in France so people could replicate it elsewhere, right?"

I nodded and swallowed a mouthful of potato.

"But a lot has changed between 1980 and today," he said.

"Agreed. Fewer mullets," I said.

"I meant cardiovascular care. Back then we didn't even give people aspirin, did we? What was the name of that study you always talk about? The one with the zodiac signs?"

"ISIS-2." I thought of Jim and checked my phone but there was no message from him.

"Things have come a long way," Alexi said.

They had. Aspirin wasn't the only innovation of the past 30 years. Many of the medications we have today simply didn't exist back then. The blood pressure pills of the day were either not that effective or poorly tolerated, and physicians were generally unenthusiastic about prescribing them. A blood pressure of 180 used to be considered mild hypertension whereas now people panic with a blood pressure of 150. High cholesterol was generally ignored, and diabetes was treated with insulin. And people smoked more, for good measure. When you consider how we practiced medicine back then, it's not surprising heart disease was so common.

It's amazing sometimes how quickly things can change. Alexi and I reflected on the past and took turns coming up with examples. Even within our own relatively short careers, things we learned 10 years ago as trainees had been overtaken by new discoveries. Looking back 30 years is jarring.

"It makes you wonder if 30 years from now, two doctors will be sitting at this table and making fun of the bonkers stuff we're doing today," Alexi said.

"Almost certainly. Our personal ridicule is the price we have to pay for progress." I can apparently become quite eloquent when drinking.

"So here's the thing. The landscape in cardiology is completely different," Alexi said.

"True."

"And if you look at differences between France and the U.S. in terms of heart disease, you can probably explain that away by the fact these are very different populations of people living very different lives within very different health care systems that take very different care of their populations."

"No argument from me."

"That's refreshing. But you still have to explain why red wine looks like it's good for you if you only drink one or two glasses per day." He was likely referring to the many studies that show that alcohol has this U-shaped association. The risk dips a bit when you go from zero drinks to one or two but then starts going up again the more you drink. It sort of looks like a 'u' or a 'j' when you draw it out. People always bring up the U-shaped curve. Repeatedly. But it's not real. The difficult part is explaining why it happens and then convincing people that it's not true. I don't always succeed. "It's . . . complicated."

"Life often is."

"First things first. It may not be red wine that's protective."

"My fault for making you drink so much. What do you think we've been talking about all this time?"

"I mean, it's not clear if this only applies to red wine or any type of alcohol."

"Isn't the theory that red wine has antioxidants in it?"

That's the usual placeholder argument people use. For a while, antioxidants were touted as the cure to everything. Free radicals became the root cause of all disease and antioxidants the solution. Medicine is never that simple though.

"What was the one that was so popular a while back?" Alexi searched his memory. "Resveratrol!"

"It was de rigueur for a while. But it didn't hold up."

"If memory serves, and I freely admit I haven't been keeping up with the research, people looked at it to prevent cancer and slow down the aging process."

"Sure. In mice."

"Not in humans?"

"Studies in humans have been very disappointing. The InCHIANTI study —"

"Greatest name ever by the way."

"By far. They found no link between resveratrol and heart disease, cancer, or anything else."

"Maybe the dose was too low."

"Possible. But the more substantive problem is wine isn't a very good source of resveratrol. There's very little of it in white wine and even red wine has fairly low amounts."

"How much do you have to drink to get your daily allowance? Bear in mind that a high number will be the more popular answer."

It was a difficult question to answer because there's no clear consensus on how much resveratrol you need. Studies have usually used around one gram as the standard dose, but some have used more. Either way, a fairly large dose is needed.

"To get enough resveratrol inside you, you'd have to drink a lot of wine," I said.

"I'm willing to make the sacrifice. Depends on the vintage though. I have standards."

"To get one gram of resveratrol in your bloodstream you probably have to drink somewhere around 500 liters of red wine per day."

"I haven't studied hepatology in a while, but I'm pretty sure that will kill you."

There is some debate in the literature about how much resveratrol you can get from wine and how much you need on a daily basis, so there's a range of estimates about how much red wine you'd need to drink. But even the most conservative estimates come out to 50 liters per day. That is marginally better but just as fatal. In practice, you can't drink enough red wine on a daily basis to get a useful amount of resveratrol, which means it's not resveratrol that makes red wine look protective because you can't drink enough of it without dying a horrible, painful death from acute liver failure.

"I've never actually seen someone die of liver failure," Alexi said.

"It's not pretty."

"But there are many antioxidants in red wine," he said. He held his glass up to the light but whether to inspect or admire it, I couldn't tell. "Maybe something else is responsible."

"Maybe. But at this point, you're just playing antioxidant whack-a-mole. If resveratrol doesn't work, you bring out another candidate. When you debunk that, then everybody says, 'Oh no, it was this antioxidant all along.' And you just keep going through this cycle where you hold up and then discount an endless line of polyphenols as possible silver bullets."

"Easy tiger," Alexi said.

We laughed. I tend to get animated when debunking misinformation.

"But something has to make red wine different. Otherwise, why is it good for your heart while white wine isn't?"

"Interestingly enough . . ."

"I always know you're going to say something annoying when you say, 'Interestingly enough.'"

"I'm glad you've been paying attention. Interestingly enough, the protective effect isn't limited to red wine. You see it with other forms of alcohol too."

"That is the great tragedy of science isn't it," Alexi said somewhat sadly. "Beautiful theories being slain by ugly facts. So maybe it's alcohol itself that's beneficial?"

"But alcohol is just a sugar molecule. Drink enough of it and it makes you gain weight. It's hard to see how something that makes you gain weight can be good for you."

But Alexi was unwilling to give up on the notion that moderate drinking is good for you. Part of the problem is that few people know what "moderate" drinking means. Or, more accurately, it means different things to different people. Ask someone how large a standard drink is and you get inaccurate estimates. Alexi's answer was telling.

"Depends on the size of the glass I'm holding at any given moment of time. For example, a standard drink is *this* much." He held up his glass. "Woops, hold on." He put the glass down, filled it and thrust it forward. "Sorry, *this* much."

"I think you've overestimated a bit."

"So what's the official definition of a standard drink?"

"That's the problem, there's no universal definition. It varies from country to country."

"But a glass of wine is more or less the same the world over."

"Alright, how many standard glasses are in one bottle?"

Alexi held up the bottle and hefted it in his hand and eyed it appraisingly. "Convention is to split a bottle between two people."

"A bottle has five servings in it."

"You might need to do that math again. I'm almost certain you're wrong."

"Standard bottles are 750 milliliters. One serving of alcohol is 13.6 grams. Since most wine is 12 percent alcohol by volume, one serving would be just under 150 milliliters. So there are five servings in one bottle."

"I'm not doing the math," he said.

"You have a calculator on your phone."

"You better keep those observations to yourself when you're at a dinner party. It won't make you very popular. Standard dinner party rules are one bottle for two people."

"That may be the rule of thumb for being a good host, but that's not what 'moderate' drinking means." I try to avoid using air quotes whenever possible, but sometimes the situation calls for it. Alexi snorted, but I wasn't fazed because I had research to back up my argument. They've done studies where they ask people to pour out a serving of alcohol and people invariably pour too much.

Alexi was still dubious, so I told him to look carefully at his glass of wine. It had a barely perceptible white line on the side. Easy to miss but there if you look for it. While not in universal use, they have become more common in recent years. Restaurants use them so their servers know how much wine to pour into a glass.

Alexi peered at it and frowned. "I'll be damned. I never noticed that before. I always thought some restaurants were just cheap."

I was tempted to press my point now that I had the advantage, but Alexi cut me off.

"You're being a bit of a buzzkill. Don't you realize people just want you to tell them what they want to hear? People like living in semi-denial."

"Doesn't the truth matter?"

"Only if the truth is at the bottom of this bottle." He topped up our glasses.

"My issue is the guidelines are too high. Even if people learned that a glass of wine is smaller than they think it is, one to two drinks per day, if you do the math —"

"I'm not getting my phone out. I've told you that."

I wanted to point out that we'd won a math competition together back in high school. "One to two drinks per day is still too high."

"What's the threshold then?"

"Two per week."

I don't know what Alexi said next. I assumed it was a curse word in Italian.

The exact hinge point where risk starts going up is debatable. Some analyses have found the mortality starts going up at around 100 grams of alcohol per week, roughly one drink per day. Others have found a lower threshold of around two drinks per week. Either way, the common belief that two per day is not only safe but good for you prevails among the general population despite evidence to the contrary.

"But does the U-shaped association still hold?" Alexi asked. His argument was that even if the threshold for safe drinking was lower than previously thought, the presence of a U-shaped curve in the data would suggest that complete abstinence is undesirable. The desire for small indulgences can lead to motivated reasoning.

"In some analyses, it does."

"Doesn't that prove the point?"

"No, because there's a more plausible explanation. The 'benefit' from low-level drinking" — I was using air quotes again and I was doing it unapologetically at this point — "is very small, inconsistent, and limited only to heart attacks. You don't see it for strokes, heart failure, or the overall risk of death. And the reason for that is simple."

"Don't say statistics."

"Yes! Statistics!" I feel like if he didn't want to hear me talk about statistics, he shouldn't have started the conversation. Although I wasn't really sure at this point who had started the conversation to be honest.

"Is there anything I can do to stop you from going on a lengthy monologue about this?" he asked.

"Order us some coffee when they come to clear the plates. I'm going to the bathroom."

"Tiramisu for dessert?" Alexi asked.

Now, part of me acknowledges the importance of leading by example and adopting a healthy lifestyle to prevent cardiovascular disease. But another part of me couldn't care less. That part tends to get more of a say when I'm out with friends. "Sure. Okay." But after sober second thought, I reconsidered my rash decision. "Ask if they have millefeuille instead."

I went to the bathroom, and as I was washing my hands (because one does not completely abandon all public health measures even after a bottle of wine) I felt my phone buzz in my pocket. It was Jim replying to my message. *We're out for dinner. Going well.*

I was pleased. I made a mental note to tell Casey tomorrow and was surprised that such a thing would have occurred to me. I went back to the table and told Alexi that my unintentional matchmaking had borne fruit.

"Ask them where they went for dinner. And ask them what wine they ordered. That will tell us how things are going."

"Why would that matter?"

"If they're drinking expensive wine at a fancy restaurant that means something. Trust me."

Alexi didn't explain what, if anything, it actually meant, but I was curious. Jim answered back almost immediately. The name of the wine meant nothing to me, but Alexi nodded approvingly and said things looked positive. It also turned out that they were eating at a restaurant not far away. We looked it up online and it was an upscale vegetarian restaurant.

"Jim is pulling out all the stops," Alexi said.

"Fortune favors the bold."

"You should take that to heart yourself."

I didn't really understand what he meant by that.

"Tell them to come over. I kind of want to meet them now. I have this image in my head and I want to see if it matches up."

"I think they may have other things on their mind tonight."

"They can stop by the conference tomorrow then. We can see them between sessions."

That seemed somewhat less intrusive, so I sent Jim a text telling him to pop in if he was free. I told him to bring Katie along too. Perhaps that would give him an excuse to see her again before she flew back home.

"So?" Alexi asked.

"We'll have to wait for an answer."

"Let's wish them well then." He drained the last bit of wine left in his glass. "We can also order a dessert wine if you like."

"I think we've had enough alcohol-wise."

"You're really that worried about your health?" he asked.

"I'm worried about finding my hotel room."

"That's why they write the room number on the little envelope they put your keycard in."

"That is admittedly very helpful."

Jim didn't look like he was going to answer my text right away, so I put the phone back in my pocket. "So do you want me to tell you why the U-shaped association isn't real?" I asked.

"Can I stop you?"

"Not really. Not without the threat of violence."

"Tempting . . . but no. Go ahead until the waiter comes."

The explanation was reverse causality, although it's sometimes called the sick-quitter as well. The basic idea is simple enough. If you do a study showing that A causes B, then reverse causality is when in fact B causes A. Mathematically, there's nothing that distinguishes the two. You differentiate cause and effect based on what came first because cause must precede effect.

Sometimes it's easy to tease out and sometimes less so. There's a famous example with smoking that often illustrates the point.

"Do you know who Ronald Fisher is?" I asked.

"You can safely assume that I do not."

"He was the guy who invented the p-value."

"Ah." Thankfully, Alexi knew the basics of what a p-value is, and I was spared the torture of explaining it to him like I did with Jim.

"I'm always surprised more people don't know his name given how often we use p-values in medical research."

"I expect that Grogg the caveman is equally puzzled that he hasn't been immortalized for his discovery that you could squeeze a cow's udder and drink the stuff that came out."

"I don't think that's how that happened."

"Why exactly are you bringing up Ronald Fisher?"

"Because he didn't think that smoking caused lung cancer."

"He may not be the statistical genius you're making him out to be then."

Problem is, even geniuses get it wrong sometimes. There's even a name for it. It's called the Nobel disease and it happens when Nobel laureates, undeniably experts in their field, start believing bizarre things. Linus Pauling went down a strange rabbit hole regarding vitamin C and thought it could cure almost anything. Charles Richet won a Nobel for his work on anaphylaxis and allergic reactions. He also believed in ghosts.

Fisher, for his part, refused to believe that cigarettes caused lung cancer despite the mounting evidence that they did. He argued that lung cancer made people smoke more. Mathematically, his argument was valid. A simple statistical association can work both ways. But a modicum of common sense demonstrates that people start smoking well before they develop lung cancer. The former obviously causes the latter.

But sometimes it's not as clear. If you analyzed hospital data, you'd probably find people admitted to hospital with depression or other psychiatric problems were more likely to be smokers. Does that mean smoking causes depression or does that mean people with depression smoke more? It's the second. But unlike the example with Fisher, it's a bit more ambiguous. Which is the cause and which is the effect?

If you were researching the impact of sedentary behavior on health outcomes, you might send out questionnaires to find out if people who were more active were less likely to have chronic diseases like heart disease or cancer. They almost certainly would. But given that your questionnaire is only a snapshot in time, the onus is then on you as a researcher to figure out if a sedentary lifestyle made people sick, or if people with chronic diseases do fewer activities because they don't feel well. Both are possible. It depends what came first. But you might not be able to pick that up on a single questionnaire.

"This is giving me a headache," Alexi complained. I thought it might have more to do with all the wine we had and said as much. He reverted back to Italian profanity, so I let the matter drop.

"Here's a more common example. You've heard of the obesity paradox?"

He nodded. "It's the idea that being a little bit overweight protects you from things like heart disease and cancer. It reassures me that many of my life choices aren't that self-destructive."

"The obesity paradox isn't real. It's reverse causation."

"Why do you have to ruin everything?"

"It's my nature. And I wish it were true. But it's not. Being a little bit overweight doesn't make cancer outcomes better. Remember cancer patients lose weight as the disease progresses."

"Cachexia," Alexi said. *Cachexia* was the medical term for weight loss due to a chronic illness and was frequently seen in cancer and heart failure patients. This is a common observation. Patients with more aggressive and more advanced cancers start losing weight as the disease progresses, whereas patients with more benign disease usually maintain their baseline weight. If you took a cross section of all the cancer patients in a clinic, the ones who weighed less would have worse outcomes because they had the more aggressive disease. The ones who weighed more would have more benign diseases. Being thin doesn't make cancer worse. Bad cancers make you lose weight. Alexi had seen more than his fair share of patients wasting away with advanced disease and the mood darkened a bit.

"I see," he said simply. "I've never thought about it that way, but it's something we all see all the time."

"Same thing applies with obesity and heart failure. People talk about the obesity paradox. But obesity doesn't protect you when you have heart failure. It's that the patients with more mild disease just don't lose weight."

"I guess I should have been inherently skeptical of the obesity paradox given that it has the word *paradox* right there in the title."

"It is a pretty big red flag."

"So the French paradox —"

"Same thing. Reverse causation. Abstaining from alcohol doesn't make you sick. People who are sick abstain from alcohol. That's why it's also called the sick-quitter effect."

There are many reasons why someone would be told to stop drinking, or at least cut down. Liver disease and cirrhosis are the obvious examples. But alcohol also increases your blood pressure. Excessive amounts can weaken the heart muscle and cause heart failure. There is a lot of sugar in alcohol and cutting back has obvious benefits for diabetics. Reduce your alcohol consumption, and you will likely lose weight as well. It's not an accident that we refer to it as beer belly rather than celebrity belly. There have also been studies showing that cutting alcohol out of your diet can decrease the burden of arrhythmias like atrial fibrillation.

"There's a lot of benefits to cutting alcohol out of your diet," I said.

"Didn't stop you from drinking the wine tonight."

"The point I'm trying to make is people drink alcohol for all sorts of reasons —"

"Agreed. I, for example, drink to make people more interesting at dinner parties."

"But people also abstain from alcohol for all sorts of reasons. Cultural reasons, religious reasons, but also for medical reasons. That's the point."

I took a napkin and sketched out a graph I use in my lectures. It was from a study that looked at the relationship between alcohol consumption and heart failure. I think it beautifully highlights the differences between never drinkers and former drinkers.

"If you lump all non-drinkers together, you get a U-shaped curve. But when you separate the never drinkers from former drinkers you realize why that happens. The former drinkers are the ones at higher risk, and they drive the association."

Ideally, you want researchers to separate out former drinkers and never drinkers in the same way that most research papers differentiate between current smokers, former smokers, and never smokers. But

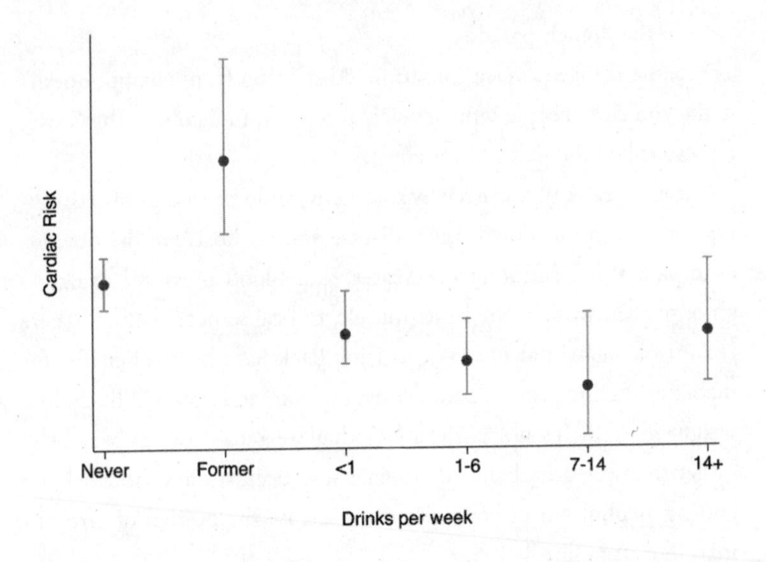

not everyone makes this distinction in alcohol research. They should, but they don't.

There are statistical designs that you can use to overcome the problem of reverse causality in alcohol research. You could track alcohol use over time, you could exclude people who had medical problems at baseline, and you can try to adjust statistically for variables that can distort the association between alcohol and cardiovascular health. But one of the more interesting techniques in recent years has been Mendelian randomization. I mentioned this to Alexi, but he didn't share my enthusiasm.

"It's a fascinating area of research," I said.

"Fascinating to you I suppose."

"I stand by my statement."

"And they're doing . . . what exactly in these studies?"

"Well, the basic premise is that your genetics don't change over time."

"Unless you get bitten by a radioactive spider."

"For those of us who aren't Spiderman, genetics is basically a random act. It depends which sperm meets which egg."

"You shouldn't be having sex with random people. That's how syphilis happens."

"Be serious."

"Make me."

In our younger days, this would have been the point when we started wrestling.

Mendelian randomization is not a simple thing to explain, but at its core is the idea that you can mimic a randomized trial using genetics. At conception, whether an embryo gets version A or version B of a gene is a random act. The key is to pick a gene that only affects one thing, like alcohol consumption. As long as the gene doesn't also affect things like blood pressure, diabetes risk, or propensity to smoke, then people with version A or version B should be identical in every other respect. Any differences between the two groups must therefore be due to their genetic differences and by extension their alcohol intake.

Alexi raised a number of objections. You have to make a series of assumptions for Mendelian randomization studies to work. The gene in question should have a strong link to the factor you want to investigate. Weakly associated genes don't work as well. You're also assuming that genes only affect one thing, which isn't always true. But despite these limitations, Mendelian randomization can be a very useful tool, especially when you can't do a proper randomized trial.

I told him about a large Mendelian randomization study done in China. It had genetic data on half a million people and looked at the association between alcohol and stroke. What made the study interesting was they analyzed the data two ways. They did a conventional analysis using questionnaires to ask people how much alcohol they drank per week and compared those results to a genetic analysis.

The conventional epidemiologic analysis showed a U-shaped association for alcohol consumption, with moderate drinking (100 grams of alcohol per week) being associated with a lower risk for stroke. But the genetic analysis showed no U-shaped curve. The risk went up the more you drank.

I'd reached the climax of the story, but Alexi said nothing. There was a moment of silence before he said, "Oh, you're done?"

"I am."

"But how do you know that the genetic analysis is correct and the usual way is wrong? Maybe it's the reverse."

"Doubtful. People who drink alcohol are different from people who don't in so many ways. Their income is higher. Their diet is different. There's no way to account for all this properly. I think most people would agree the genetic analysis is closer to the truth."

"The obvious way to resolve this issue is with a proper study, a proper randomized trial."

"That would be ideal, but it's probably not going to happen."

"Why not?"

Before I could answer, the waiter returned. The kitchen was backed up, and they had run out of the desserts we'd ordered. They still had fruit and crème brûlée if we were willing to accept substitutes. Alexi looked at me, and I shrugged, but then a thought occurred to me. "I wonder if Angelo's is still open."

He smiled. "Let's go find out." He asked the waiter for the bill. The waiter returned, we settled our account and walked into the night.

"Which way is it? Do you remember?" Alexi asked.

"I'm pretty sure this way." I nodded to the left.

"When was the last time we were there?"

"I can't remember. Not since we were in training."

"I hope they're still open."

We walked in silence, and when we turned the corner, the street looked familiar. We went another few blocks and walked past the restaurant Jim and Katie were eating at.

"Do you think they're still in there?" he asked.

"It looks like it's emptying out." The restaurant had a large street-front window that let you look in. But it seemed like most customers had come and gone and the evening was winding down. I checked my phone out of curiosity and saw Jim had answered.

They were no longer at the restaurant, but he suggested we could do breakfast tomorrow. That suited me just fine, but Alexi had an early morning meeting he couldn't miss. I told him to text me if his meeting finished early, and if not, I'd find him after my talk and we could have coffee.

"I'll meet your barista at least if I can't meet Jim and Katie."

"She's not *my* barista, and at some point, we should probably listen to at least some of the conference sessions."

"We have to put our priorities in order," Alexi said, but he made no effort to explain what that order should be. "By the way, you never said why a randomized trial on alcohol will never happen."

"There was going to be one. It was called the MACH trial."

"Hold on, let me guess. Moderate . . . alcohol . . . in cardiovascular . . ."

"Moderate Alcohol and Cardiovascular Health. They were going to randomize 7,800 men and women to zero or one drink per day and see if it affected rates of heart disease."

"Doesn't 7,800 people seem a little on the small side?"

"It does, a little bit. That was one of the criticisms."

"There were others?"

"Oh goodness yes. The trial was only going to last six years, which is short if you want to look at cancer outcomes because it takes many years for cancers to grow and develop. Also, when they designed the study to look for cardiac issues, they removed heart failure as one of the possible endpoints."

"Seems strange they would leave it out."

"Unclear if it was an oversight or deliberate. The NIH commissioned a report into what was going on with the trial and they found staff were in regular contact with industry representatives."

"You mean with the liquor industry?"

"Turns out the liquor industry had put up a lot of money to fund the trial. That amount of input and influence was casting doubt on" — I had to choose my words carefully — "the trial's impartiality."

Alexi is often blunter than me. "You mean they were stacking the deck. They didn't want a study to say their product makes people sick."

"Final report said industry involvement in planning the study went 'beyond the norm.' The NIH pulled the plug because concerns about the study design 'cast doubt on its ultimate credibility.' They said something like that."

"I can't believe this stuff keeps happening."

I thought about my conversation with Jim and Katie about the salt study on prisoners.

"I know. Controversies like this are not helpful. It's really got to stop."

"It's a shame because a study like that, if you did it properly and didn't get into bed with industry, would actually be really useful and settle the issue. We've been talking about alcohol and heart disease for 40 years. It would be nice to get a definitive conclusion."

"It's a statistical artifact," I said. "It doesn't hold up when you use genetic data and there are so many situations where alcohol is harmful. Even if there's a small benefit in preventing heart attacks, it's outweighed by the increase in cancer, heart failure, arrhythmias, and everything else."

"It doesn't mean a randomized trial will *never* happen."

"I think the MACH story poisoned the well. Plus the idea of red wine being heart-healthy is so widespread, it's one of these myths that will never die."

"I keep forgetting we humans are not rational animals. We're rationalizing animals."

I laughed at that. I was going to have to borrow it.

"Do you think ice cream would help?" Alexi asked me. We'd arrived outside Angelo's, and even at this hour there was a small crowd of people sitting on the benches outside the store eating soft serve, sundaes, and frozen yogurt of all kinds.

"It can't hurt."

"I wouldn't get too worked up about the MACH trial. The fact they investigated and pulled the plug means the system worked."

"I suppose. But it would be nice if it didn't happen in the first place."

"I know," Alexi said. "But we can fix the world tomorrow. Tonight, let's just have some ice cream. I'm buying."

MYTH #6

Chocolate Is Health Food

Ice cream may not be able to fix all the world's problems, but it comes close. Alexi and I sat at a small picnic table outside Angelo's enjoying our dessert. When we got there, a few patrons were milling about and sitting at the tables outside, but the crowd thinned as the evening stretched on. Ten years ago, when we'd attended the same conference with other residents in our cohort, we'd also ended up at Angelo's at midnight. The place hadn't changed.

"Do you still speak with anybody from back then?" he asked.

"A few people. Emails and messages, stuff like that."

Alexi nodded. He kept working at his ice cream cone, so we sat in companionable silence.

"I've more or less lost touch with everyone," he said. "Any major news?"

I went through a mental catalog of everyone in our training program. "Nick got married."

"That I knew," Alexi said. "Facebook."

"Right." We went back and forth, calling up names from the past and comparing notes about what had happened to whom. It was so easy to lose touch with people. We'd spent years together, worked 100-hour weeks, and spent our 20s seeing each other more than our own families. But now I could go months without thinking of them.

"I never would have thought you'd be the lone holdout in the marriage department," he said. "I'm actually kind of surprised I'm getting married before you are."

I shrugged.

"Why did you order vanilla?" he asked.

I didn't really have a good answer. "I guess I just like vanilla," I said. "I like the taste."

"But vanilla ice cream doesn't taste like anything."

"Untrue. It tastes like vanilla."

"It's the plainest of flavors. Chocolate is better."

He did seem to be enjoying the chocolate. I thought of telling him that the great benefit of living in a free and democratic country is that people can enjoy whatever ice cream flavor they want. I was pretty sure that wasn't what Solon of Athens was thinking about when he first suggested the idea, but it was a pleasant side benefit. That and not being subjected to Soviet-style political purges.

I decided to respond with a more conciliatory position. "I got my chocolate anyway."

One of the perks of Angelo's is that every order came with a piece of milk chocolate or dark chocolate. If you ordered a sundae or one of their fancier dishes, it would be wedged into the ice cream like a garnish. But if, like us, you stuck with the ice cream cone that made the place famous, then it came as a separately wrapped bite-sized piece you collected at the cash register. I held mine up as proof.

"But you got milk chocolate," Alexi said.

"I like it better than dark chocolate. Dark chocolate is too bitter."

"Dark chocolate is better for you though."

I made a noncommittal sound that was halfway between a grunt and a mumble.

"Are you going to ruin chocolate now too?"

"To be fair, you keep bringing this stuff up," I told him.

"Don't you remember that Nobel study?"

He was referring to a study published in the *New England Journal of Medicine*. It was a brief report, but it got a lot of press coverage. It was a simple analysis, which probably helped with its broad appeal. It found that countries where people eat a lot of chocolate also produce more Nobel Laureates. "I remember," I said.

"You sound skeptical."

"The fact that you ordered chocolate ice cream isn't going to help you win a Nobel Prize."

"And why not?!"

"Because you don't do medical research." His disdain for the number crunching that occupies most of my day was legendary and well-documented.

"True. That would be a rate-limiting step."

As a pure clinician, he was possibly the smarter of the two of us. He was definitely the richer one. "Keep trying," I said. "I believe in you. You can do anything if you put your mind to it."

"Was that a real study?"

"What do you mean a real study?"

Alexi told me about an article he'd read by a journalist who'd attempted to prank medical publishers. I'd heard of it at the time, but I'd forgotten the details. It was a documentary being produced for German television about how easy it was to turn junk science into news headlines. To prove the point, they threw together a diet study, of sorts. They recruited 16 people and split them into a control group, a low-carb diet group, and a low-carb plus an extra 1.5 ounces of chocolate group. The chocolate group lost more weight by the study's end, three weeks later. They wrote up a paper and got it published. Then various news outlets picked it up.

Alexi and I had a long debate about whether it was a "real" study. On the one hand, it did really happen. Real people were recruited, and they did follow the described protocol. The data was real, and the numbers weren't fudged in any way, at least according to the documentary's production crew. My counterargument was more subtle. This study was in service of a documentary about how easy it was to get junk science published and have it generate flawed news headlines.

The study wasn't an exercise in creative fiction, but it had multiple problems that I was only too happy to detail for my friend. It was too small, too short in duration, and the authors committed the cardinal sin of analyzing every possible outcome under the sun to find something newsworthy that would entice bloggers and media outlets. They weren't being dishonest or self-serving. They were trying to prove a point. It was published in one of the many journals that will publish anything for a fee, and it faced no scientific scrutiny or peer review. I told Alexi that the study might be real. I just wasn't sure the results were.

"To be fair, it's not as if this was featured on CNN or anything," he said.

It's true. It wasn't. But it was highlighted on a number of websites and different media outlets around the world. And, I pointed out, their budget was small and their timeline short. With enough PR resources, they probably could have whipped up more coverage. But that wasn't their objective. The point wasn't to prank people. The point of the documentary was to expose a problem with how we talk about food research. We see an interesting story and a catchy headline, and we don't kick the tires of the underlying research all that hard.

"Out of curiosity," Alexi asked, "did the study or the news coverage make any distinction between milk chocolate and dark chocolate?"

"The study didn't, and most of the news outlets just copy-and-pasted the press release the authors put out."

"Isn't there some value in making a distinction between dark chocolate and milk chocolate since dark chocolate has more antioxidants in it?"

"Maybe," I said. "There are antioxidants like flavonols in the cocoa bean, and the assumption is that they would help prevent or treat disease."

"I'm sensing a big 'but' coming."

"But" — What can I say? I know how to play to an audience — "the amount of flavonols in processed chocolate is not the same as in raw cocoa beans."

"I'm guessing it's less."

"Very much so. The manufacturing process destroys many of the flavonols in chocolate."

"Life is often unfair like that."

This is one of the many problems in this area of research. Are you talking about an extract from the cocoa bean, or are you talking about chocolate sold in stores? The distinction is often blurred in lay press coverage.

But Alexi didn't want to let his original point go. He argued that the flavonol content in dark chocolate is higher than in milk chocolate.

"It's true that dark chocolate has a higher cocoa content, but they both have less flavonols than the original cocoa bean they were made from." For good measure, I added, "And as much as people like to claim that they like dark chocolate, more people eat milk chocolate by a wide margin."

Alexi conceded that a lot of people don't like the bitter taste. Most people don't even realize that chocolate is naturally bitter and only tastes good because it's loaded with fat and sugar to mask its taste.

"So if I've understood you correctly, if we could get people to switch from milk chocolate to dark chocolate they would be healthier."

"Just by cutting out all that extra sugar, they probably would be." I could tell Alexi was prepared to declare victory. I raised my hand. "But I'm not sure those extra antioxidants would do anything. Plus, people don't eat chocolate to be healthy. People eat chocolate because it tastes good. You can't convince people to eat something higher in cocoa if it's so bitter that it becomes unpalatable."

Alexi put his arms on my shoulders and looked me in the eye. With a straight face, he said, "If anyone can do it, *you* can. I believe in you."

I brushed away his arms. "You're such a jerk sometimes."

He laughed. "Come on, let's get up and walk a bit. It's a nice night, and I'm not ready to head back to the hotel." He got up and tossed his balled-up napkin into the trash. At three feet, it wasn't a difficult shot, but he was proud and expected me to be impressed.

"I missed my calling." He said it wistfully.

"For what?" I asked him. "The NBA? How were you going to grow the extra six inches you need to be competitive?"

"I'll be one of those short and fast NBA players you always talk about."

For all his apparent disinterest in medical research, Alexi did absorb stuff. Angelo's had closed, and we were the only two left. But despite the late hour, people were out walking about and enjoying the unseasonably warm evening. We walked in silence for several blocks.

"I was just thinking of something, and it tied into what we said about red wine over dinner," Alexi said.

I murmured encouragement.

"It doesn't matter if there are antioxidants in chocolate. It only matters if eating chocolate prevents disease."

It was a good point, and I told him so. There isn't a lot of clinical research involving actual chocolate and hard clinical outcomes, but there is some. "There was the Danish study," I said. "Did you review that one?"

"You're going to have to be more specific. There's been more than one study from Denmark over the course of human history."

He was, to some degree, an infuriating person. "This was from the Danish Diet, Cancer, and Health Study. They sent people questionnaires about their diet."

"Food questionnaires?" Alexi asked. "Aren't you always going on about how food questionnaires are inaccurate and prone to . . . what did you call it?" He searched for the words and snapped his fingers. "Recall bias."

These are the moments when you realize you've made a difference in the lives of others. "Yes, recall bias. Exactly right."

"Sorry, I interrupted you. They sent out food questionnaires and —"

"They found that the people who ate chocolate had a lower risk of atrial fibrillation."

"And why do we think that chocolate can help prevent arrhythmias?"

"The explanations do get a little bit hand-wavy, but the idea is that chocolate's antioxidants suppress cardiac inflammation and something something something less arrhythmia."

"I think you glossed over some of the details there."

I had. But clarity on the mechanism of action was often lacking. Many people assume the antioxidants in chocolate would mediate its supposed health benefits, but Alexi pointed out that we'd said the same thing about the antioxidants in red wine. On that point, he was quite right. "They looked at dark chocolate?" he asked.

"Most questionnaires don't make the distinction between milk chocolate and dark chocolate, so you can't tease out that difference."

"Seems important given that the entire premise of the argument is based on the idea that dark chocolate is healthier."

It was indeed a shortcoming, but the authors defended it by pointing out that in Denmark, even though most people eat milk chocolate, not dark chocolate, milk chocolate has more cocoa than in other countries. Milk chocolate in Denmark has minimum 30 percent cocoa compared to 10 percent cocoa in the United States. The popularity of milk chocolate over dark chocolate isn't the only issue that muddies the water. Not all the studies agree. Some show a benefit and some do not. Data from the Women's Health Study showed a benefit while data from the Physicians' Health Study didn't.

"Maybe there's a sex difference, and it only works for women and not men."

There are few people in this world I can talk shop with. And Alexi, for all his pretense, raised good points and asked good questions whenever we got into these discussions. "Possible, but the data from Denmark didn't show a difference between men and women. So you're just left with some positive studies and some negative studies." Whenever you have many different studies from many different sources with varying results, there needs to be a review to try to come to some sort of unifying conclusion. There is a review like that for chocolate but it was looking at blood pressure as an outcome, rather than arrhythmias.

"Why would they think the antioxidants would lower blood pressure?" Alexi asked.

"They don't. It's the magnesium."

"Is there a lot of magnesium in chocolate?"

"More than most foods, I guess."

"I'm a bit more removed from this than you," Alexi said, "but doesn't magnesium only lower blood pressure if you inject it intravenously?"

He was right, but when he gets on a roll, he has a habit of cutting me off.

He stopped walking and grabbed my arm to make his point. "Plus, if magnesium is good for you, we can just give people magnesium. We don't have to coat it in chocolate for people to take it. That's how you get dogs to take medicine."

I laughed. That's one advantage vets had over us.

"Chocolate seems like a particularly unhealthy magnesium delivery system if you ask me," he said.

Nobody had, but I agreed with him again. "And its ability to lower blood pressure is pretty modest. If you summarize all the randomized studies of chocolate or cocoa, there was barely any effect on blood pressure. On average only a 1.8-point reduction," I said.

"That's not very much," Alexi said.

"No, it's not. It's pretty trivial and not worth all the extra calories, fat, and sugar you're getting with the chocolate."

"All the fat, sugar, and calories you're getting with chocolate is probably the very reason why it *doesn't* lower blood pressure."

He was probably right. I told him so, partly so he could let go of my arm, partly so we could start walking again. I told him I'd blow his mind with another aspect of chocolate research that often goes overlooked.

"Hit me," Alexi said.

I proceeded to do exactly that.

He rubbed his shoulder. "I meant, tell me what this other problem is."

"The question you should be asking is, 'Who paid for the study?'"

"Must be government grants. I can't see pharma paying for this."

"Not the pharmaceutical industry, no. The chocolate industry."

"You're telling me Big Chocolate is a thing?"

Chocolate companies spend millions on research. They fund studies, set up research programs, and organize conferences. Mars actually has a research department called Mars Symbioscience and an organization called the Mars Center for Cocoa Health Science. Alexi listened but he didn't seem especially bothered by these revelations.

"But these are private industries. If they want to do their own in-house research —"

"They also funded a research department at UC Davis to study cocoa."

"That's harder to overlook."

Industry sponsorship casts doubt on the validity of research. It doesn't mean that it's introducing bias into the research, but the potential for bias is there. If you go back to the Cochrane review on chocolate and high blood pressure, some of the studies included in the meta-analysis were sponsored by industry and had authors that were employees of the company. They tended to show a larger benefit and excluding them from the analysis tended to weaken the association between chocolate and blood pressure. The authors concluded there was a possible commercial bias in the research.

Alexi made the case that it wasn't completely definitive, which was true. But there does seem to be a trend for industry-funded studies to skew positive. Vox had an article about chocolate companies and their influence on research. They looked through 100 chocolate studies that were funded or supported by industry, and 98 of them came back positive.

"You'd think we'd be better about drawing a line between private companies and scientific research," he said.

I didn't have a particularly good answer to that, so I said nothing. I was about to say something profound about the importance of research ethics when I realized where we were. "We've been here before."

"You mean metaphorically or are you saying we've been walking in a circle?"

"No, I mean the last time we were here at the conference as residents."

"That was 10 years ago. Are you telling me you actually remember a specific street corner you walked past a decade ago?"

I ignored the question and took my bearings. If Angelo's was behind us, the park should be a few blocks to our left. I walked and Alexi followed.

"Do you want to tell me where we're going?" he asked.

"It'll be more fun if I just show you."

"More fun for who?"

"Well, me, obviously. Now if memory serves, we turn right at the next corner, and it should be one block over."

"If memory serves? You realize you have an app on your phone that gives you directions to places, right?"

"I don't like to rely on technology."

"The other day you emailed me to say you just bought a $5,000 computer to analyze medical statistics."

"Yes, well . . . that's different. Anyway, we're here."

Alexi looked around. "I'm unclear as to why you wanted to show me a seesaw."

"I wanted to revisit the site of the most epic game of hide and seek ever played by medical residents at a conference."

"That was pretty epic." He closed his eyes, and I imagined that like me he was replaying the evening in his head. "That game got intense. How long did you hide for?"

"I think about 45 minutes give or take."

"That was ludicrously long for you to stay hidden. I remember we started to get worried that maybe something happened to you. There was some serious discussion at one point about whether we should call the police and file a missing person's report."

"I take hide and seek very seriously."

"What exactly were you doing all that time?"

"Eve and I started talking. At first, we were just staring up at the sky looking at the stars, and she started asking me if I knew any of the constellations."

"Do you actually know the names of the constellations?"

"I'm not sure I know the names of *all* of them, but definitely the big ones."

"Do you find that information useful?"

"It comes up in conversation more often than you'd think."

"Useful with the ladies, is it?"

Alexi liked to needle me when he was bored. "I don't know," I said honestly. "It only worked that once."

"You know, as a general rule, talking isn't the best strategy during hide and seek."

"Nobody found us, though, did they?"

"That might have had more to do with the fact that everyone was three sheets to the wind."

"That was certainly a contributing factor."

Alexi looked uncertain, like he wanted to say something but wasn't sure. "Do you speak to her family at all?"

It helped if I took a deep breath before answering. "No, we slowly drifted apart. It was probably my fault. I never knew what to say to them, and I stopped making the effort after she was gone. I figured it was probably easier for them to stop seeing me as well."

"It's not wrong to talk about her."

"It's not easy either."

We said nothing for a long while. Just stood there at the edge of the park.

"Tell me a happy memory," Alexi said. "Tell me why you came out of your hiding place."

"Eve wanted ice cream. She said all perfect evenings have to end with ice cream. So she went to find the rest of you."

"I didn't know it was her idea to go to Angelo's."

"It was the hotel's recommendation. She was good at that. Striking up conversations with people she'd just met, I mean."

"Do you know she challenged me to an ice cream eating contest that night?"

"Really?"

"Yes. I declined."

"That's for the best. She would have won." You wouldn't have thought it from her petite frame, but it was true. "She also liked chocolate ice cream," I added. "She never understood why I liked vanilla either."

"Chocolate is objectively better."

"That's not what the word *objectively* means."

"Whatever. I feel like we should be able to settle this issue once and for all. It would just take a large randomized trial. I'm sure we could convince some large government agency to fund a multimillion-dollar study."

"It's already been done."

"There's a large randomized trial on chocolate ice cream?"

"Well, on chocolate anyway."

"It wasn't sponsored by a chocolate company, right?"

I didn't say anything.

"Right?"

"The study was sponsored in part by Mars, the candy company."

"Damn it," Alexi said as he rubbed his temples. "We really do live in the stupidest timeline."

"I've held that opinion for a while now."

"Did this study get quashed too like the alcohol one?"

"No, it went forward and got published. To their credit, the authors were committed to publishing the study regardless of what the results were."

"I don't remember hearing anything about a large study on chocolate."

The study didn't make as much of a splash as it should have. It was called the COSMOS study, short for the Cocoa Supplement and Multivitamin Outcome Study. Alexi applauded the acronym even if the letters didn't quite map out perfectly. As the name implied, it tested a cocoa extract rather than commercially available chocolate. It also had a parallel study arm that tested a multivitamin supplement. Prior to its publication, everyone talked about the study as if it would settle the whole chocolate controversy. But when it was published and the cocoa extract showed no major benefit in terms of cardiovascular disease, there was little publicity. I saw it only because I was looking for it, and to this day I am puzzled by the underwhelming response.

Alexi wasn't surprised. "It's disappointing but not entirely surprising. If it were positive, I expect the chocolate companies would be screaming it from the rooftops."

"One presumes. Being able to tell people chocolate is healthy sounds like it would be key to their marketing campaign."

"Was there anything else interesting about the study?"

"Apart from the cardiovascular stuff, they also looked at whether the cocoa extract prevented cancer."

"Did it?"

"No."

"It sounds as if the day the COSMOS study was published was not a good day for the chocolate industry."

"Somehow, I think they'll be okay."

As we talked, we walked over to a soccer field and stretched out on the pitch in defiance of grass stains and the damp. It was, just as it was 10 years ago, a great place to stare up at a cloudless star-filled sky. There was no way to completely get away from the light pollution of the city, but we could see stars we hadn't seen or thought to look for in years. We lay there on the grass for a while, staring up at the sky. We'd likely pay for it in the morning, but at that moment, the interplay of alcohol and ice cream left us perfectly relaxed and heedless of the aches and pains tomorrow would bring.

Alexi rolled over and propped himself up on his elbow. "I just thought of something."

"You mean how we need to be up in about four hours for the conference." I was remarkably unconcerned.

"No, I was thinking about the study linking chocolate to Nobel Prizes. Given that the evidence for chocolate is so shaky, how could something like that get published in what is maybe the top medical journal in the world."

"It was probably a joke that got out of hand. When you read the article, it does seem like it's supposed to be satire."

"I don't think everyone was in on the joke. A lot of media reports took it very seriously."

"I know. But when you read the article from start to finish, you realize that there's a lot of winking at the camera."

"For example?"

"Well, in the disclosure section, it says that Dr. Messerli, the author of the piece —"

"He wasn't funded by a chocolate company, was he?"

"No, the disclosure section said, 'Dr. Messerli reports regular daily chocolate consumption, mostly but not exclusively in the form of Lindt's dark varieties.'"

Alexi screwed up his face. "I don't think his chocolate preferences are relevant."

"They aren't. In a serious paper, you'd just say that you have no disclosures or no conflicts of interest. You wouldn't take it as an excuse to tell people what kind of chocolate you like." I called up the paper on my phone to show him another example. When you read the paper, Sweden seems to be an outlier and doesn't fit the overall pattern with more Nobel Prizes than you'd expect. I read Alexi the explanation in the paper: "One cannot quite escape the notion that either the Nobel Committee in Stockholm has some inherent patriotic bias when assessing the candidates for these awards or, perhaps, that the Swedes are particularly sensitive to chocolate, and even minuscule amounts greatly enhance their cognition."

"That's an odd way to phrase things."

"I'm pretty sure it was supposed to be a joke. There's another part where he says maybe the explanation for these findings is that winning a Nobel sets off a country-wide celebration of widespread chocolate consumption."

"The more you tell me about this paper, the more convinced I am that he's trolling us."

That was my first impression too. When I first heard about the paper, I was certain it was one of those light-hearted, tongue-in-cheek, satirical articles you see in the Christmas issues of medical journals. But the coverage seemed to imply that it was serious research. The problem was most articles never interviewed Messerli and never realized he was trying to make a point about bad science.

"So it was all fake?"

"It wasn't made up if that's what you mean. The data is real. If you plot out the data points, you do get a straight line."

It wasn't a perfectly straight line. Some countries like Sweden were outliers, but the overall graph was fairly linear.

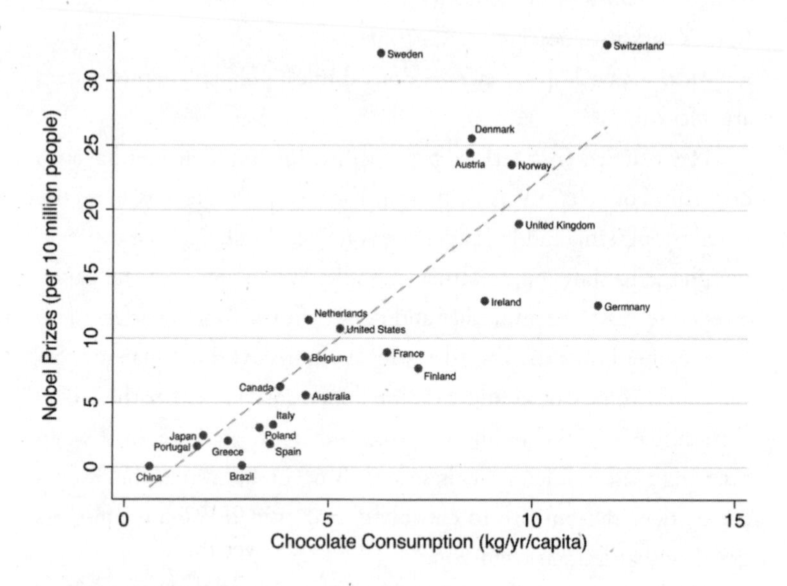

"But what was the source of the data?" Alexi asked.

"Wikipedia mostly and a few other websites to get the chocolate statistics for each country."

"Did Wikipedia become an authoritative source of medical information when I wasn't looking?"

"I hope not."

"I'm starting to see why you think this was all satire."

I had to admit to Alexi that I was in doubt for a while. But one night, when I couldn't sleep and was mindlessly searching the internet, I stumbled on an interview Messerli gave where he explained his reasoning for doing the research.

One evening, he was in a hotel room in Kathmandu and saw a study about flavonols improving scores on cognitive tests. He decided to look into the issue. He figured Nobel Prizes were probably a good sign of overall intelligence, and he knew chocolate has flavonols. A few quick Google searches later, he had the chocolate consumption per capita of each country and the number of Nobel Prizes. He plotted the data out and was amazed by how strong the correlation was.

At first. In an interview he gave to Reuters, he called the association meaningless and the whole idea absurd. He also said he was trying to make a point about how people can misuse data to draw faulty conclusions. The same news article interviewed Eric Cornell, who won the Nobel for physics in 2001. He obviously thought it was a joke because he was quoted as saying: "I attribute essentially all my success to the very large amount of chocolate that I consume. Personally, I feel that milk chocolate makes you stupid. Now dark chocolate is the way to go. It's one thing if you want like a medicine or chemistry Nobel Prize, okay, but if you want a physics Nobel Prize it pretty much has got to be dark chocolate."

Alexi laughed. But he had some sympathy for Cornell, who obviously thought he was making an off-hand comment that wasn't going to be taken seriously. Instead, it was sent out over the Reuters wire service and got reprinted in numerous articles across the internet.

"Somebody should tell journalists that different departments mock each other all the time."

"It's true. I make fun of surgeons continuously."

Alexi's eyes narrowed slightly. "Oh really? How so?"

"What do you call two surgeons trying to read an ECG?"

He shrugged.

"A double-blind study."

"Okay. How many cardiologists does it take to screw in a lightbulb?"

I didn't know so I waited for the punchline. I assumed it would be a large number.

"Six," Alexi said. "One to screw in the lightbulb and five to say how much better they could have done it."

All good comedy is based in truth. We traded barbs for a while until Alexi remembered what had gotten us onto this tangent.

"I wonder if the moral of the story is don't try to be funny when giving an interview."

"That's very cynical. I don't think we can fault the journalist. The original Reuters article was very clear."

When you read the whole article, Cornell does go on to say that he was joking and that he didn't think the association was real. He makes the very good point that richer countries eat more chocolate, because it's a luxury item, and richer countries spend more on medical research and therefore are more likely to win Nobel Prizes. It was a very astute observation but the only thing that got reposted was the original tongue-in-cheek quote about milk chocolate making you stupid.

The BBC radio show *More or Less* tried to interview Cornell afterward to clarify his statement and he sent them a written response that read: "I deeply regret the rash remarks I made to the media. We scientists should strive to maintain objective neutrality and refrain from declaring our affiliation either with milk chocolate or with dark chocolate. Now I ask that the media kindly respect my family's privacy in this difficult time."

Alexi let out a low whistle. "That does not sound like a man who enjoyed seeing his name in print."

"Imagine winning a Nobel Prize and all people can ask you about is what kind of chocolate you eat."

"It does sound exhausting."

"The thing that really bothered me about this study though was that it committed the ecological fallacy."

"You're going to have to remind me what that is."

"The ecological fallacy is when you mistakenly assume that what's true for the group is also true for the individual."

"You might have to unpack that one a bit," Alexi said.

"Let me give you a concrete example. Is it better to be tall or short?"

"Always better to be tall unless you're flying on a plane."

"I mean in terms of heart disease risk."

"I don't think it matters all that much."

Amazingly it does. There has been a lot of research looking at whether height can predict cardiac disease. Alexi wondered aloud whether that was the best use of research funds, but the minds of the people who approve research funding were often inscrutable. I argued there was some value to the research because height was being used as a surrogate for growth hormone, or IGF-1, which affects both height and possibly heart disease risk as well. But Alexi had zero interest in this field of research, so he stifled a yawn and told me to proceed with whatever point I was trying to make.

"The point is research linking height to things like heart disease, cancer, or lifespan is interesting because —"

"I once again feel compelled to remind you that we have different definitions of the word *interesting*."

"I stand by my use of the word."

I genuinely believe it's interesting because something weird happens when you look at people from different countries. If you take people in Denmark and compare them to people in Italy, you find that on average Danes are taller and have higher rates of heart disease. Alexi pointed out that such things would be interesting only if you were willing to overlook the many other differences between Italians and Danes. That was a fair enough comment. But things get really interesting when you compare countries and individuals within those countries. Danes are taller than Italians, and Denmark has more heart disease. But if you look at people within those countries, taller people have less heart disease.

I thought I'd be able to explain it better with a graph. I tried to show one to Alexi but he took off his shoe and threw it at me and told me to just enjoy the evening.

It's a visually impressive relationship when you see it for the first time. Northern Europeans are taller and have more heart disease than southern Europeans.

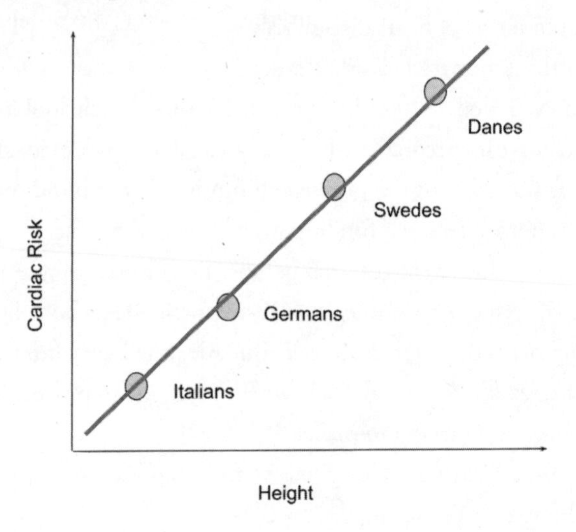

But if you plot out the data points from individual people, you get a different picture. Within each country, the risk trends downward in taller people.

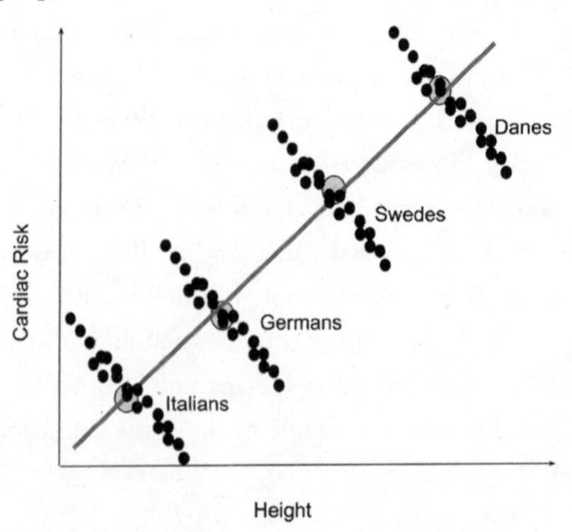

In other words, the relationship flips in the opposite direction when you look at individuals rather than the population average. It seems counterintuitive but is surprisingly common. It's called the ecological fallacy.

Alexi asked the obvious question: Which was the right way to analyze the data? Should you look at the aggregate population average or the individual-level data? I told him the answer was that it depends on what you're interested in studying. It's not wrong to compare countries, or states, or counties, or cities, or groups of people of some kind. But if you want to draw conclusions about how something affects individuals, then you need individual-level data. Otherwise, you could draw the wrong conclusion from your data.

"There's plenty of other famous examples of the ecological fallacy," I said.

"If you're going to keep using the word *famous* like that, you have to start using air quotes."

"Back in the 19th century, Émile Durkheim wrote a famous book called *Le Suicide*."

"Upbeat title."

"It looked at the link between religion and suicide rates in Prussia."

"That keeps me up at night."

"Hey, shut up. So he found that provinces with a higher proportion of Protestants had greater suicide rates and those with more Catholics had less."

"Most people would have been too afraid to bring religion into a discussion like this. Good for you for not having boundaries."

Now it was my turn to throw *my* shoe at *him*. At least it would have been, but when I sat up to get at my shoe, the knots wouldn't cooperate, so I flopped back down on the ground. Which was just as well because I wasn't keen to go looking for it afterward. After I caught my breath, I finished my analogy. It's a classic example of the ecological fallacy and often used to teach the concept to students. Although provinces or countries with more Protestants might have more suicides, if you look at the suicide rates of individuals, there isn't much of a difference. People have since reanalyzed Durkheim's data and found that when you look at the individual level, the difference between Catholics and Protestants is much less pronounced.

"You know very weird and obscure things," Alexi said. "Nothing that directly applies to medicine, mind you."

"I do have a medical example."

"Of course you do."

"You've heard that fatty food increases the risk of breast cancer?"

"I've heard people say it. But it's not actually true."

"It's not. Or at least the studies have been largely inconsistent and mostly negative. But when you look at *countries*, it definitely looks like it does."

"I know you think you're making sense but you're not."

"It's simple." At least it usually is. I was tempted to just go to sleep there lying on the ground looking at the stars, but I wanted to get this final thought out. "Countries with high-fat diets have more breast cancer. That's true. But those countries are usually developed nations with older populations where breast cancer rates are high to begin with."

Alexi grunted but he kept his eyes closed so I thought I might be losing him.

"But you're assuming the national average applies equally to every individual. If you just look at the summary data of a country, you haven't shown that the women who got breast cancer actually ate a high-fat diet," I said.

"I'm guessing the same thing applies to chocolate. If you're trying to prove that chocolate boosts your mental performance . . ."

"You really need individual-level data. *Countries* don't win Nobel Prizes and *countries* don't eat chocolate. *People* win Nobel Prizes and *people* eat chocolate."

"And you can't establish that the people who won the Nobel Prizes were the ones who were eating the chocolate," Alexi said.

That was the fundamental point. Satire or not, the Nobel chocolate study in the *New England Journal of Medicine* couldn't prove what it claimed even if it wanted to. Alexi seemed convinced. Convinced or asleep, it was hard to tell. But I think he got the point. I figured I could probably close my eyes for a bit.

A moment later, Alexi, possibly inspired by a dream, rolled over. "A randomized trial would obviously be helpful here. The COSMOS study didn't look at brain performance, did it?"

"It did. There was a sub-study of COSMOS called COSMOS-Mind that looked at cognitive performance. But it didn't show any benefit. The cocoa supplement did nothing."

"This has all been vaguely disappointing," Alexi said. "Like most of our get-togethers."

"It's not my fault chocolate isn't health food."

"Well, I have to blame someone. It's not as if I can take personal responsibility for my actions."

"No, of course not. That's completely out of fashion."

Alexi agreed. Neither of us felt the need to say anything. The only thing that prodded us into movement was the pinkish hue on the horizon.

"What's that thing you used to say?" Alexi asked me. "From Greek mythology? About the dawn?"

"When the child of the morning, rosy-fingered Dawn, appeared."

"Yeah, that's it. Rosy-fingered Dawn has appeared, and it's time for us to get back to the hotel."

"It's going to be a rough day."

"It's not the first time the two of us have gone without sleep."

"But it's been quite a while. We're not 20 anymore."

"I am according to the Julian calendar." Alexi thought for a moment and turned serious. "Stop talking like an old man. You're not even 40 yet."

"Almost."

"You're not there yet. You still have a lot of living to do. You have to make up for lost time."

I wasn't sure what he meant by that. "I'm going to need a lot of coffee tomorrow."

"Good. I can meet your barista friend."

We picked ourselves up. The sun peeked its head above the horizon, which made it much easier for Alexi to find his shoe. Properly shod, we started walking back to the hotel. I remembered something. "Jim and Katie are coming by the hotel for breakfast." I checked the time on my phone. "In under two hours." I felt like crying.

"We get to see if your matchmaking paid off!"

"That wasn't really my plan."

"Don't you remember the old saying: life is what happens when you're busy making other plans."

"I don't base my life on things I read in fortune cookies."

"Don't be so negative. It's a brand new day. Interesting things are going to happen. I can feel it in my bones."

I looked back at the sunrise. It was a beautiful start to the day.

MYTH #7

Breakfast's the Most Important Meal of the Day

I got 90 minutes of sleep before I had to get up to meet Jim and Katie for breakfast. It wasn't much, but it was better than nothing, and it wasn't the first time I'd gone without sleep. After a quick shower, shave, and change of clothes, I headed downstairs. Jim and Katie had considerately agreed to meet me at my hotel since I had to give my presentation later.

As I walked across the reception area, I saw them seated in the restaurant. I was disappointed. I'd wanted to see if they arrived together. Jim looked much the same since one dark suit looks much like another.

Katie on the other hand looked very different. For the flight, she'd had her hair back in a ponytail and had been decked out in sneakers, yoga pants, and a t-shirt, an obvious concession to comfort over style. Over the course of our conversation about red meat and salt, I'd formed the impression that she was a rational person, and no rational person wears heels on an airplane. Now, though, she was dressed for important meetings in a business suit that was stylish, tailored, and expensive. Her shoes probably cost more than her airline ticket.

We exchanged hellos. "I didn't realize you were staying here," she said. "My hotel is just down the street."

As impressed as I was by her sartorial style, her comment threw me for a loop.

She must have seen my puzzlement. "We could have shared a cab from the airport. Jim said you were going a different way, that's why you couldn't ride with us."

"Yes." I said it slowly as I tried to think of what to say next. I looked at Jim, and he looked at me with just a hint of panic, pulling me deeper into our fraternal conspiracy. "Well . . . I had to go meet a colleague uptown." That seemed eminently plausible.

Katie seemed satisfied with that explanation.

"What's good here?" Jim asked.

As generic topic-changers go, the classics are classics for a reason. We perused the breakfast menu for a moment. Jim exclaimed that $4.50 extra to add bacon to your meal seemed excessive. That drew a narrow-eyed stare from Katie.

Jim coughed. "I probably wouldn't have ordered the bacon anyway." He looked down at his menu to avoid eye contact. "Maybe the pancakes." That seemed like the safest bet.

The waiter came, poured water, and took our orders. Katie ordered cinnamon oatmeal, I went with an omelet, and Jim at the last minute changed his mind and ordered two eggs over easy.

"Very good," said the waiter as he scribbled. "Will that be with ham, bacon, or sausage?"

Jim froze.

"Sir?" the waiter prompted. "Which would you prefer?"

Jim opened and closed his mouth.

"Just get the bacon, Jim," Katie said.

The waiter took Katie's instruction as definitive. He scribbled something down on his pad and left us. Jim wiped away a little bit of sweat from his hairline and I, taking pity on him, tried to deflect Katie's attention. "You like cinnamon?" Talking about what people ordered is always a good idea when you have nothing else to say.

"Yes, I even put it in my coffee when I can find it," she said.

"I've never heard of anyone putting cinnamon in their coffee before."

"Not that rare. It's common enough that they put it out in coffee shops, so people must use it."

Fair point. "Is cinnamon oatmeal your go-to breakfast then?"

The waiter arrived with a pot of coffee and refilled our water glasses. I added some milk to my coffee and noticed Katie took hers black. Jim took the milk from me and reached for the sugar. He paused and picked up and sipped his coffee instead.

"I actually rarely eat breakfast," Katie said. "I often have early morning meetings. I usually get by with just coffee." She tapped the side of her cup for emphasis as she raised it to her lips.

"But isn't breakfast the most important meal of the day?" Jim asked.

"Do you know where that comes from?" I asked.

Jim shook his head, and even Katie looked like she wanted the answer.

"It was a marketing slogan."

"No!" Jim said. "That can't be right. Everyone says that. My doctor told me that."

"It was a *very* successful marketing campaign," I said. "It's gotten to the point where it's ingrained in the public consciousness, and nobody remembers where it comes from. But basically it was the Kellogg brothers."

"The guys who invented corn flakes?" Jim seemed astonished.

"Those guys, yes." But as we were talking about it, something shook loose from the deep recesses of my memory. "But I think it was someone else who first came up with the idea."

"I have it," Katie said. She read off her phone. "The first person to suggest that breakfast was the most important meal of the day was dietitian Lenna Cooper who wrote it in an article for *Good Health*, a magazine published by the Battle Creek Sanitarium, which was run by Dr. John Harvey Kellogg."

I explained to Jim that in that era, a sanitarium didn't mean an asylum for the mentally ill. It was essentially the 19th-century equivalent of a modern health spa. Sadly, not everyone enjoys the history of medicine and philology as much as I do. Jim was keen to argue the point about breakfast being the most important meal. He didn't think any PR campaign could be persuasive enough to change people's eating habits wholesale.

"It did with bacon," I said.

"What do you mean bacon?" Jim looked a little worried.

"Bacon only became popular as a breakfast food in the 1920s or so because of a marketing campaign. It was spearheaded by Sigmund Freud's nephew, I think. He was a big ad man and did a bunch of the big PR campaigns."

Jim had the look children get upon discovering that Santa Claus isn't real. He looked to Katie. She looked up from her phone and nodded.

"So it's a conspiracy. We're being manipulated by big business. They're telling us to do stuff not because it's healthy but because it helps their bottom line. I don't know what to believe anymore," he said.

I was worried that if I didn't pull Jim out of his tailspin he'd start thinking the moon landing was faked by the time we finished breakfast. "It wasn't a grand conspiracy. It's not even a secret. A bacon company hired a marketing firm to promote their product, and they succeeded. People like the taste of bacon. It probably didn't take much prodding to get them to eat more of it."

"That's true," Jim agreed. "Bacon does taste amazing."

Katie closed her eyes and rubbed the bridge of her nose.

"The point is, you can eat whatever you want for breakfast. There aren't rules. Habits change over time. Our habits, for good or ill, are manipulated by marketing. What seems like normal breakfast food to us would have seemed strange one hundred years ago."

"That's fascinating," Jim said. He sipped his coffee. "Bacon and cereal were invented to make a quick buck."

I'd attended a lecture once on the history of breakfast as a concept. Historically, most people didn't eat breakfast early in the morning. Their morning meal would have been around lunchtime. But that changed as people moved from farm life to industrial work where jobs started at a fixed time every morning. What we eat and when we eat has been a moving target throughout human history.

Cereal is a classic breakfast food today, but that wasn't always the case. Cereal, or rather the cold flakey cereal that we're used to, was invented by the Kellogg brothers, though they spent years arguing about who deserved most of the credit. And depending on who you want to believe, it was invented either as a way to treat digestive issues in their sanitarium patients or as a way to curb people's romantic passions.

It turned out John Harvey Kellogg, the director of the sanitarium, had some very unorthodox ideas about food. His big innovation was encouraging people to eat bland foods to help with digestion, a very de rigueur concept at the time. He also thought that spicy or overly stimulating foods could excite the libido. Kellogg thought his new cold cereal would help calm his patients' passions. Lust and passion were, in his view, detrimental to people's well-being.

At least, that's the popular version of the story. It might be more folklore than reality. Kellogg never marketed his cereal as a sexual suppressant. There's indirect evidence from things he wrote at other points during his career that he probably believed it was, to some degree. But the invention of cereal was initially about making an easy-to-digest breakfast food for the sanitarium patients. The sex thing may or may not have come into it later.

"This all seems sketchy," Katie said.

"Much of the 19th and early 20th century was sketchy as far as medicine is concerned," I said.

"If breakfast isn't the most important meal, I feel somewhat vindicated in skipping it," Katie said.

"So why spend so much time and money on school breakfast programs?" Jim asked.

"It sort of depends on what you mean by important," I said. "School breakfast programs do seem to help students perform better, especially if kids aren't getting breakfast at home."

The science and benefits of breakfast are clear when you talk about hungry children in school. The problem is when people transpose the data from school programs into other contexts. There's a difference between ensuring proper nutrition in school-age children and saying that eating waffles will make you better at your job. Jim shared a personal anecdote. He couldn't skip breakfast. If he did, he couldn't concentrate at work. He'd get a headache, his stomach growled all day, and he would be intensely irritable. He also told me there was a term for that. *Hangry*.

You learn something new every day.

The waiter arrived with our food, and Katie stirred her oatmeal to let it cool while I added some pepper to my omelet. Jim dug into his breakfast platter.

"So which of the three of us is eating healthier right now?" Katie asked. She blew on the oatmeal.

"Probably you."

She smiled, pleased that she'd one-upped both me and Jim. He'd gotten into the heart of his eggs and was using his toast corners to dab up the yolk spilled across his plate. When he was done, he circled back to his earlier point. He was pretty sure he'd seen or heard about research that skipping meals was bad for you. He probably meant that when you spread out your meals over the day, you have more consistent blood sugar levels and avoid large spikes and fluctuations. I told him that it probably did matter to some degree if you were diabetic, especially if you were a diabetic who needed insulin.

"So does skipping breakfast cause any problems in somebody like me who's not diabetic?" Katie asked.

"There was one study I remember linking skipping breakfast to heart disease."

"How would skipping breakfast do that?"

"That's up for debate. There was an analysis from the Health Professional Follow-up Study. A large cohort of doctors, dentists, pharmacists, and vets was followed over time and given food questionnaires every so often to ask them what they eat."

"But wouldn't the problem be the same one we mentioned before? Recall bias?"

I was impressed with Katie. Most people don't pick this stuff up as quickly.

"Very possibly," I said. "Although when you're asking people *whether* they eat breakfast or not, that tends to remain fairly constant over time.

That's not to say that people don't change their habits as they get older, but it usually doesn't change from one day to the next."

Katie agreed. "Breakfast people are breakfast people." She looked at Jim, who was spearing home fries with his fork.

"And eating breakfast is a big enough part of your daily routine," I said. "It's unlikely people would remember that wrong."

"Are you suggesting skipping breakfast might be putting me at risk for a heart attack?" she asked.

Jim nudged Katie with his elbow. "In that case, you can have some of my potatoes." He pointed at his plate. It was hard to tell if he was being funny or serious.

I cleared my throat. "Before you start changing your morning routine, the study had some caveats. The people who skipped breakfast were different than the people who didn't. Younger, which probably worked in their favor. But they were also more likely to smoke, drink, work full time, and exercise less."

"That makes sense," Katie said. She told us she skipped breakfast because she started work early and didn't feel like force-feeding herself at 5:30 a.m. It was logical to assume that the people who skipped breakfast worked long hours or did shift work and simply didn't have the time to eat at regular intervals. They were at higher risk because they had other unhealthy lifestyle habits.

"Can't you adjust for all that?" Jim said with a mouthful of egg. "Adjust for those differences statistically, I mean."

It wasn't an unreasonable question. "Up to a point, yes. But you'll never account for those differences completely no matter how hard you try."

"Why not?" Jim asked.

At least I thought that's what he asked. He didn't swallow before talking. "You can't quantify human behavior perfectly. Think of all the lifestyle choices you make daily that might affect your health."

Jim thought about it and said I had a point.

"That study did say skipping breakfast increased heart attacks. But when you adjust for things like diabetes, cholesterol, high blood pressure, and weight, the link tends to weaken and the association melts away."

"So it probably wasn't so much skipping breakfast but more those other risk factors that were making the biggest difference," Katie said.

"Very likely."

"Wouldn't skipping breakfast help you lose weight though?" she asked. "That would obviously help you be healthier."

"I thought it was the opposite," Jim said. "I thought skipping breakfast made you *gain* weight. Isn't the whole idea that by skipping breakfast you end up snacking more throughout the day?"

"That's 100 percent wrong." Katie dug in. "You skip meals, which makes your body burn more fat and so you lose weight. That's the logic behind intermittent fasting."

I thought some diplomacy was in order. "Both theories have some merit —"

"Because you skip breakfast but then end up having a muffin with your morning coffee to compensate," Jim said. "Which is basically like having cake, so you end up having more calories."

"True," I said.

"I'm not so sure about that," Katie said. "It's not as if I start going on a hunger rampage, cramming Danishes in my mouth."

In fact, it was hard to look at Katie and see how skipping breakfast could lead to weight gain. Katie was slim. She'd be the first to go in a famine but that scenario was unlikely in the short term. "The evidence for breakfast as a weight loss tool leaves a lot to be desired."

Jim wasn't pleased. In 2019, a meta-analysis in the *British Medical Journal* reviewed all the randomized studies looking at the effects of breakfast on weight gain. The authors found eating breakfast had very little effect. Contrary to Jim's hypothesis, skipping breakfast resulted in a small weight decrease. But the difference was less than one pound. Essentially there was no difference.

Katie crowed, but I was more cautious. "There was a lot of inconsistency in the data," I said, "and many of the studies had quality issues. We should probably say the data is inconclusive."

"Doesn't intermittent fasting have research propping it up?" she asked.

"Depends what you mean by intermittent fasting. Originally it meant the 5:2 diet where you only dieted on some days of the week. But then it became about not eating at certain times of the day."

The 5:2 diet centered on feast days, where you ate whatever you wanted, and fast days, a bit of a misnomer since you didn't actually fast on those days. You just ate much less than usual. The 5:2 ratio was a popular diet at one time, but there were many subsequent versions. Katie had never actually heard of the 5:2 diet. She was more familiar with time-restricted eating, where you were supposed to fast only at specific times. You either stopped eating at midnight till noon the next day. Or from 8 p.m. until the following lunchtime. Or from 6 p.m. to 8 a.m. There were many different versions, but the basic idea was the same. Stop eating and force your body to burn fat.

"It makes sense to me," Katie said. "But it can be challenging to stick with."

"It's not particularly hard to fast for 12 hours when you're sleeping during eight of them," Jim said. "It's kind of hard to eat while you're sleeping."

Katie tried hard not to smile.

"Very true," I said. "For all its rules, intermittent fasting just boils down to skipping a meal. Sometimes that meal is breakfast, and sometimes that meal is dinner."

Katie said she had dinner at about 7 or 8 p.m., skipped breakfast, and then had lunch at her desk at around noon. "That's sixteen hours between meals, isn't it?"

Jim tried to do the math in his head.

I put him out of his misery and agreed.

"You see," Katie said, "I've been doing intermittent fasting for years. I just started doing it before it was trendy or had a name."

"The real question though is does it help you lose weight?" Jim asked.

"It's a tough question to answer because intermittent fasting has changed so much," I said. "Recent studies like the TREAT trial —"

"Who gets to pick the names of these trials?" Jim asked. "I'm genuinely curious."

"The authors," I told him. "For many researchers, it's their only creative outlet."

"Do the acronyms mean anything beyond being clever?"

"Not really. They're supposed to be acronyms of the study title, but these days they rarely are. They're just a convenient shorthand. Rather than quote you a whole reference, I can just tell you the TREAT trial was negative, and you'll know which study I mean."

"Well, other doctors will. Not me."

"Fair enough."

"So eating breakfast or deliberately skipping breakfast doesn't affect your weight?" Katie asked. "That surprises me."

"Not in the grand scheme of things. If it has any benefit, it's probably small and inconsistent."

Since I'd been doing most of the talking I was behind the others in finishing my breakfast. Jim had polished off his eggs and was making short order of the last few pieces of fruit. Katie had made significant progress on her oatmeal, but I'd barely made any headway on my omelet. Jim eyed my plate, but hadn't yet come up with a convincing and discreet way to ask if I intended to finish that.

Katie was also looking at my plate. "I'm amazed the two of you can eat such large breakfasts."

"It's a skill," Jim said.

"I'd need a nap after breakfast if I ate like that every day."

"To be fair, no one's telling you that you can't," Jim said. "You could still order some eggs if you wanted to." He scanned the restaurant for our waiter.

Katie snorted. "Ha! No. I never eat eggs. I never had them growing up. It's just not part of my routine."

"Why not?"

"My parents heard eggs were bad for you because of the cholesterol, so they didn't let me or my brother have any."

"You didn't turn to eggs as a sign of teenage rebellion?" Jim asked.

"No. I did what any self-respecting teenage girl would do. I started smoking so I could look cool. But I never took to it, and I gave it up by the time I started college."

"Thank goodness for small mercies," I said.

"So you've never had eggs?" Jim asked.

Katie shook her head. "Afraid not."

Jim let out a low, long whistle of disbelief.

"If it's any help, the thinking about cholesterol in eggs has changed," I said.

Jim nodded. "Cholesterol is good for you now, and all the stuff people used to say about cutting fat and cholesterol out of our diet is wrong."

"That might be overstating things," I said. "But you're right. Current guidelines have relaxed restrictions about cholesterol."

"I thought it was all completely debunked," Jim said. "Now we know the problem isn't fat but sugar. I think I saw a book or article about that. It said we should avoid sugar and eat more fat, cream, butter, eggs, and bacon."

"So someone told you eating bacon was healthy and you believed them?" Katie asked.

"I did. It made a lot of sense."

Katie looked like she was going to say something but didn't.

I jumped in. "There's no shortage of websites telling you fat is good. But there's also no shortage of websites telling you fat is bad. It's a problem when they say both things simultaneously." I pulled out my phone and showed him a screenshot a friend had sent me. I keep it for situations just like this.

Jim handed my phone back. "You see, this is why people don't trust scientists."

I shrugged. The criticism had some validity.

"So which is it?" Jim asked. He leaned back, likely anticipating a long-winded answer.

"Working under the assumption that neither of you wants a one-hour lecture" — they both nodded — "the answer depends on the

The Telegraph @Telegraph · Aug 17, 2018

Low-carb, high-fat diets could knock years off lifespan, 25-year study suggests

telegraph.co.uk

Low-carb, high-fat diets could knock years off lifespan, 25-year study ...

Eating a diet which is low in carbohydrates could knock years off lifespan, a 25 year study suggests.

SOURCE: TWITTER

The Telegraph
@Telegraph

Low-fat diet could kill you, major study shows

telegraph.co.uk

Low-fat diet could kill you, major study shows

Low-fat diets could raise the risk of early death by almost one quarter, a major study has found.

SOURCE: TWITTER

type of fat. You have trans fats, saturated fats, mono-unsaturated fats, poly-unsaturated fats, and those are all different. So probably safer to say that you should avoid unhealthy sources of fat."

"So avoid trans fats in pre-processed foods?" Jim asked.

"That's a good first step."

"But what about eggs? I like eating a good hearty breakfast of eggs and bacon to get the day started."

Katie looked to the ceiling. "You shouldn't purposefully eat *more* bacon and eggs if you want to be healthy. There's no version of the universe where that makes sense."

"But it tastes good."

"I'm sure it does. I don't actually know. But that's not the point. The point is fried bacon is not the key to longevity. Forget the cholesterol. Just eating that much processed meat isn't good for you. We talked about this."

"The issue with processed meat. Right. I forgot about that."

"You forgot bacon was made from meat?" Katie looked at me. "Back me up here."

"So there's a few things to unpack in this discussion. I think a lot of the confusion is that people have forgotten the historical context of the debate."

"What historical context?" Jim asked.

"It was really about rising rates of heart disease. Heart disease rates started going up in the 20th century, and there was a real worry that we needed to do something about it."

"Why all of a sudden?"

There was no clear consensus. Many things likely happened at once. Between 1900 and 1960 life changed dramatically for many people. More and more people moved from farms to cities. Their diets changed, and they became more sedentary. The discovery of antibiotics meant that once-fatal infections were now curable. People started living longer, and more people made it to middle age, which meant they lived long enough to develop chronic diseases like heart disease and cancer.

Jim said it was a bit depressing to think that you cured one disease only to then die of something else, but I thought a near doubling of the human lifespan was hard to argue with. And even Jim had to concede that it was weird that people used to die of pneumonia and tetanus and other things we don't think twice about anymore.

"I for one am somewhat pleased that my odds of dying during childbirth are dramatically lower than they were a century ago," Katie said.

"You see, Jim, Katie is less likely to die now because of antibiotics. She may have to deal with heart disease 30 years down the road, but it's a small price to pay."

"I'll drink to that," Jim said. He raised his cup to his lips.

You aren't supposed to toast someone with coffee, or so I've been told, but it seemed like a petty thing to point out. We could have ordered mimosas, but I still hadn't recovered from the night before. "That's the explanation. You have a combination of people living longer, smoking more, and the general shift from rural to urban life."

"When you put it that way, it's actually not much of a surprise," Jim said.

"The real question is what to do about it. And don't forget this was happening when we didn't really have any of the treatments we have today."

"I remember you saying that aspirin only started being used in the late '80s."

I nodded. "Without any effective medications, the only way to turn things around was by modifying someone's risk factors. In 1964, you had the Surgeon General's report about smoking and the beginnings of anti-smoking laws."

"That helped."

"Very much so. But more relevant to this conversation, you had the recommendations to reduce cholesterol by eating less fat."

"Which ended up being wrong," Jim said.

This was going to be complicated to explain. "Not wrong exactly, but it's more nuanced than people realize."

"But if I eat a lot of fatty food that's high in cholesterol, I won't get more plaque in my arteries," Jim said.

"Actually, you will. That was one of the first experiments linking cholesterol to heart disease. Back in 1913, Nikolai Anichkov took some rabbits and fed them large amounts of cholesterol. Then he dissected them —"

"Ugh," Katie said. "That seems pretty cruel."

"I guess it is. But if you want to prove that cholesterol is bad for you, it's much easier to prove that using animals. I mean you can't take a bunch of humans and force-feed them large amounts of purified cholesterol."

"I think you could. If there was an experiment where I had to eat bacon every day, I'd be okay with that," Jim said. He looked at Katie. "Because I like bacon."

Katie put a hand on Jim's arm. "I know, we'll work on that."

"The problem is the researchers would have to cut you open after a few months to see how your arteries were doing."

"I probably wouldn't like that part as much."

This led to a long conversation about animal research. Katie brought up the ethical issues, which are not trivial. But animal research, whether we agree with it or not, has a role to play in medical research because basic questions need to be answered in a straightforward way. You take rabbits, you feed them cholesterol, and you dissect them. The rabbits that got fed the high-cholesterol diet had more plaque in their arteries. Case closed.

It seemed logical to Jim, but he pointed out the problem with my little narrative of Anichkov's experiment. "Why didn't that settle the cholesterol controversy?"

"Nobody could repeat Anichkov's results. They tried it in dogs and rats, and nothing happened. Their cholesterol levels didn't go up and there was no build-up of plaque in their arteries. Since the experiment couldn't be replicated, it was largely ignored."

"Makes sense. What's true in rabbits isn't necessarily true in humans. After all, we're carnivores and they're vegetarians."

"Herbivores," Katie said softly.

"Right, herbivores. So it's not entirely surprising they would be more sensitive to cholesterol. They're not used to it. Our diet is much closer to dogs or wolves because they hunt and eat meat."

Katie was having none of it. "Except that's not true. We aren't like dogs. We aren't meat eaters. For most of human history, humans have primarily survived on plants, nuts, and occasionally some meat from hunting thrown in."

From what little I knew about early human history, Katie was right, and I said so, to Jim's disappointment. It's always easier to justify something if you can claim humans have been doing it for thousands of years.

"So why did the cholesterol thing work in rabbits but not dogs?" Jim asked. "I can't believe that we're more similar to bunnies than dogs." He cocked his head.

I couldn't help seeing the similarities between him and a golden retriever my friend had growing up. "Well dogs, rats, and other animals handle cholesterol differently than we do. How much do you remember from biology class in high school?"

"You can safely assume nothing," Jim said.

It worries me when people say that. It makes my job harder. I gave Jim and Katie a brief rundown of how the body handles cholesterol. When the body wants to get rid of cholesterol, it converts it into something called bile acids that are drained through the bile ducts into the intestine and removed through our feces. Jim didn't like hearing about bodily functions while we were still eating. I've become immune to such talk.

I assured him the biology portion of the talk was finished and explained that the experiment didn't work in dogs because their bodies metabolize cholesterol differently and get rid of a lot of it as bile acids. If you block some enzymes in dogs, you can see the same process that you saw in rabbits. But at the time, when the experiment couldn't be replicated in other animals, people just wrote off Anichkov and his rabbit experiment. Cholesterol research only picked up again decades later. Had Anichkov's work been recognized and accepted at the time, it would have sped up the research by years.

"So what happened?" Jim asked.

"Do you want the short version or the long version?"

"Always assume the short version."

"You had two phases of research. The epidemiology studies and the diet studies. The epidemiology studies looked at populations. On average, they found that places where people eat high-fat diets had higher rates of heart disease."

"But wasn't that the problem recently? I remember there being a huge thing about how these studies were all flawed and problematic and that the main guy behind them was manipulating the data or something. I don't really remember."

"You might be referring to Ancel Keys and the Seven Countries Study."

"That sounds right."

"Well, this is a really complicated subject —"

"Cliff Notes version," Jim said. "We have plans later." He stole a glance at Katie.

"First point. The Seven Countries Study is 70 years old, and we have almost a century's worth of newer research that is more relevant."

"I was wondering how important a study from the 1950s was in today's world."

"Frankly not very," I said. "It's sort of like rehashing the Betamax vs. VHS debate now that you have streaming services. Even the DVD player is almost obsolete."

"What's a Betamax?" Katie asked.

"It was a competitor to VHS tapes," Jim explained. "A bit before my time though. I think my parents had one."

I was tempted to say that Betamax was a bit before my time too, but I doubted either Jim or Katie cared. My main point was that old studies are like old newspapers. Historically interesting but less relevant the longer it's been. But even if you wanted to discount the Seven Countries, you have other research programs like the Framingham Heart Study, which I would argue is more important to the early history of cardiac research. Framingham was the one who identified things like cholesterol, blood pressure, smoking, and lack of exercise as risk factors for heart disease. That's actually where the term *risk factor* comes from.

"But these epidemi . . ." Jim stumbled on the word, as so many do.

"Epidemiology studies. They looked at populations as a whole."

"But you said there was a second phase to the research. The diet studies."

The history of cholesterol research isn't really divided cleanly into before and after periods. But when I explain it to people, it's useful to think about it in this way because it gives you a better understanding of why things unfolded the way they did. Jim accepted my caveats, clapped me on the shoulder, and told me we were all friends here. It was strangely touching.

Once people had teased out the role of cholesterol in promoting heart disease, they decided they had to do something about it. But in the absence of any effective medication to lower cholesterol, the only tool at their disposal was a dietary intervention. The only thing researchers could do was get people to eat less fatty food and hope it reduced their cholesterol.

"Did it?" Jim asked.

"Depends. There were a lot of different diet studies done in the 1960s. I mean there was the —"

"You aren't going to start rattling off a bunch of different study names, are you?" Jim said and smiled.

"Jim isn't a detail-oriented person," Katie said. "Just give us the take-home message."

"Fair enough. The take-home message is some studies showed a benefit and some didn't. In some studies, people's cholesterol went down a bit, but rates of heart disease didn't change much."

"That's what I thought," Jim said. "You can lower cholesterol, but it doesn't prevent heart attacks."

"That's where the argument comes from. It's decades old, mind you, but that's where it comes from."

"Don't make it wrong."

"No. But the problem, which we can now see in retrospect, is these diet studies changed cholesterol but not by that much. And if your cholesterol doesn't change by that much, obviously your cardiac risk isn't going to change by that much. That's why you got some negative trials."

"So your diet doesn't affect your cholesterol levels?"

"A bit. But for a lot of people cholesterol is genetic."

"I guess that should have been obvious."

A lot of these things should have been obvious, and they are now in retrospect. But at the time, some of the shortcomings of these early studies weren't clear. The studies were small and the change in cholesterol from dietary interventions was minimal. Some of the early studies also didn't make the distinction between total cholesterol, good cholesterol, and bad cholesterol. They didn't differentiate between trans fats and other types of fats. All of that would come later.

The studies had shortcomings and were at times contradictory. People don't always stick with their assigned diets. Compliance is hard and dropouts are common. But rather than concluding diet was, at best, an inconsistent way to lower cholesterol, people discounted the whole cholesterol theory. They said cholesterol had nothing to do with heart disease, and they believed it. Jim likened the situation to throwing out the baby with the bathwater. I agreed with him.

The hope going forward was that you could settle the issue by developing medications to lower cholesterol. That way you can do the studies in a controlled fashion and randomly assign people to the medication or a placebo. That was the next big phase of research. The medication trials of the '70s and '80s were when researchers tried medications like clofibrate and cholestyramine.

"I've never heard those names before," Jim said.

"They aren't used much these days," I said. "Mainly because they weren't that effective. They did lower cholesterol a bit though. And the risk of heart attacks was lower as a result."

"They actually sound positive," Katie said. "That would seem to argue in favor of the cholesterol hypothesis."

"But there was no difference in mortality. In the study with clofibrate, it was actually a bit higher."

"That's not good," Jim said. "That's why people say lowering your cholesterol too much is bad for you."

"People do say that. But it was only with clofibrate. The other medication, cholestyramine, was studied in the Cholesterol Primary Prevention Study —"

"The study names were less cute back then," Jim said.

"In that study, mortality was the same. Not higher. Not lower. Just the same. It was 7 percent lower in the group that got the cholesterol medication, but the margin of error was too wide to be definitive."

"So lowering cholesterol isn't dangerous?" Jim asked again. I didn't blame him. So many people have heard that lowering cholesterol is dangerous that the idea has really cemented itself in the public consciousness. I reassured him it wasn't. But some of the findings from those original studies don't make it easy to reassure people. In the Cholesterol Primary Prevention Study, overall mortality was the same, but the medication group recorded more violent deaths. That came as a surprise.

"How are there violent deaths in a medical study?" Jim asked. "I know people hate waiting in waiting rooms but . . ."

"Well, people can die of anything," I said. "All deaths are recorded during a study, and in this one more people who got the medication died of causes that were listed as accidents, homicides, or suicides."

"Why?"

"It's completely random. Random and frankly nonsensical."

Jim still seemed worried. "Could it be that lowering your cholesterol damages your brain? I've heard that's an issue. Your brain needs cholesterol to build up normal tissue and . . . you know, stuff like that."

"I've heard that too. And this study was its origin. People speculated that lowering your cholesterol affects your brain and makes it more likely you'd commit suicide. Or make bad decisions and get into an accident."

"But how do you explain getting *murdered*? Hmm?" Katie asked. "How does lower cholesterol make it more likely that you will be the *victim* of a violent crime?"

"That part never made sense," I said. "And it's pretty clear now that lowering your cholesterol isn't going to turn your brain to mush and make you a violent risk-taking sociopath."

But the somewhat unpromising results of the early diet and medication trials left an imprint on the public consciousness. They generated headlines and articles about how lowering cholesterol was dangerous and wrong. Magazine covers like one that appeared on the *Atlantic* in September 1989 talked about "The Cholesterol Myth," and that idea has never completely gone away. You can still find versions of that magazine issue online and the cover art really drove the point home. Jim looked up the magazine cover on his phone and read the headline out loud. "Lowering your cholesterol is next to impossible with diet, and often dangerous with drugs — and it won't make you live longer." He wasn't surprised people bought in to the notion that the cholesterol hypothesis was a myth.

"But you seem to be implying these arguments are no longer valid," Katie said.

"They were valid in the 1970s and 1980s, but things have changed since then. It's not the 1980s anymore."

"That's true," Jim said. "It's not cool to grow a moustache anymore."

"Thank goodness," Katie said.

"The point is science has moved on. There are newer, better medications out there."

"You mean statins?" Jim asked.

"Yes, but other medications too. They're all good at reducing cholesterol and preventing heart attacks."

"But I heard statins don't actually prevent deaths. They just pretty up the cholesterol numbers on your blood tests," Jim said.

I'd heard that often. People feel strongly about statins, which is odd because they don't have strong feelings about other cholesterol medications. But the idea that statins don't prevent cardiovascular events is a recycled talking point from the early cholesterol debate. Over the past 30 years, there have been numerous studies where statins reduced the risk of heart attack and stroke and even cardiovascular death.

Jim was skeptical because he'd heard so much to the contrary. The problem, I told him, was twofold. First, people don't die of heart attacks as much anymore, at least not from their first heart attack or stroke. They survive because now we have defibrillators to shock people back to life if they go into cardiac arrest. We unblock clogged arteries, and we have better medications to treat people. Also by reducing big risk factors like smoking and high blood pressure, people are not getting the same types of severe massive heart attacks they once did. People are simply less likely to die of a heart attack now than they were 30 years ago. That is obviously good news. But it does make research harder because if the death rate is already low, you can't make it much lower. That makes it look like the medications aren't working, when in fact they are.

Jim reluctantly agreed.

"But having more heart attacks and strokes, even if they don't immediately kill you, isn't benign. They can leave you with long-term problems," Katie said.

"That's true. It's not just about the quantity of life. If you want a better quality of life, don't have a heart attack or stroke. You're not always back to normal afterward."

This would be clearer from the research if researchers could follow patients for 20 or 30 years after a study began. That would allow us to measure the long-term impacts of non-fatal events. But long-term follow-up is logistically complicated.

Jim said it was hard for him to accept my assurance that lowering cholesterol lowers your cardiovascular risk because guidelines have done away with recommendations to cut cholesterol.

"You're right, the recommendations used to say that you should limit your daily cholesterol intake to 300 milligrams per day."

"How much is that?"

"That's the problem. Three hundred milligrams of cholesterol doesn't intuitively mean anything to anyone. People don't measure out the cholesterol in the food they eat."

"You said the same thing about salt. It's hard for people to gauge how much salt they eat because salt is in all food anyway."

"Same with cholesterol. The point is to avoid excess cholesterol, and that's why dietary guidelines are now more about recommending healthy dietary patterns rather than specific counts."

"But aren't eggs a major source of cholesterol?"

"They are. Absolutely. Most eggs have about 180 mg of cholesterol in them, give or take. They're probably the second largest source of cholesterol in most people's diets."

"What's first?"

"Meat actually. A little bit less than half of the cholesterol in our diet comes from meat, about a quarter from eggs, and the rest from everything else."

"So what you're saying," Katie looked at Jim, "is that someone who wanted to eat less cholesterol should eat less meat and more plant-based food."

"But I'm still confused about whether eating eggs is bad for you or not," Jim said.

"It is a little confusing. The data on this issue is inconsistent. There have been over a dozen studies —"

"Tick-tock," Jim said and tapped his watch.

"Okay, okay, fine. There are a lot of studies. Some suggested eating a lot of eggs increased the risk of heart attack and stroke, some showed no difference, and some showed that it was protective."

Jim brightened at that last part. "I think we should focus on those studies."

"That's not how science is supposed to work," Katie said.

Jim didn't look like he entirely agreed with that assessment. "No . . . no I suppose not."

"Why is the research so all over the place?" Katie asked.

"A lot of the stuff we already talked about. A lot of the older studies measured total fat instead of breaking it down into trans fats and unsaturated fats and all that. Also a lot of them measured total cholesterol rather than distinguishing between good and bad cholesterol."

"I see," she said. "And I guess that when it comes to eggs, you also have the problem of food questionnaires like we talked about, right? It's hard to get an accurate count of how many eggs someone eats."

"That is the obvious problem."

"And they won't necessarily count eggs in cakes and pastries and stuff you don't cook yourself. You showed us that cartoon about that."

"Also very true."

"But it's still strange that some studies would find that eating eggs *prevents* disease," Katie said.

"That's the other problem. It's the 'compared to what' problem."

"What's the 'compared to what' problem?"

"Whenever you ask, 'Is something bad for me?' you always have to ask, 'Compared to what?' If you want to know if eggs are bad for you, it depends on what you're going to have instead. Are you talking about having pancakes, or a fruit bowl with some yogurt, or skipping breakfast entirely or having a kids' cereal that's loaded with sugar —"

"Depends on whether there's a toy prize inside," Jim said.

". . . because those are different things." I had to ignore him if I was going to make my point. "Because if you just add eggs to your diet, you have to account for the fact that you're eating more calories in total than someone who skips breakfast entirely."

"I see. So it's maybe not eggs that are bad for you. It's maybe that you're just eating more."

"Now the other problem is people don't eat food in isolation. For example, people who eat eggs are more likely to eat bacon —"

"That's true," Jim said.

"So if people who eat eggs tend to eat more meat, now you have to ask yourself if they are worse off because of the eggs, or because of the meat which is also a source of cholesterol."

"Yeah, I see how that would complicate things."

"There was a study recently that generated a lot of headlines because it suggested that eating an egg a day could increase your risk of heart attack."

Jim frowned at that news.

"But, importantly, when they controlled for total cholesterol intake from other sources, the association melted away."

"That's reassuring at least," he said.

"So the issue is not so much whether you eat eggs, but what are they replacing in your diet? Because when you eat more of one thing, you invariably eat less of something else."

That is the perennial problem in food research. That *something else* will determine if eggs look good or bad. You might eat more eggs but cut out butter, fatty foods, and meat. In that case, your health status might not change. But if you eat more eggs and eat fewer fruits and vegetables, eggs are going to look harmful.

"My head hurts," Jim said. "I just want to know if I can have eggs for breakfast."

I had to laugh. "You can actually do whatever you want."

"Within reason," Katie said.

I agreed, but that wasn't my point. "Whether something is good or bad for you depends on context. My grandmother grew up in a small village in Greece and lived through World War II. When she had eggs, she would protect them jealously to give to her kids because food was scarce."

"Why was food scarce?"

"There was a famine during the war. Surprisingly, the Nazis weren't that interested in making sure everybody had enough food to eat."

"Shocking," Katie said.

"I know. In that context, having eggs probably meant the difference between survival and malnourishment. But that context isn't the context *we* currently live in. So what was true then isn't true now."

"But you do need fat and cholesterol in your diet," Jim said. "They *are* essential."

"Sure, but you don't have to get them from eggs and meat. You can get them from plant-based sources."

"See," Katie said. She elbowed him in the ribs.

"The reason the professional guidelines moved away from recommendations about cholesterol is that the situation is complex. Eating is almost like a zero-sum game. If you eat more of one thing, you end up eating less of something else. And that trade-off is very context-specific. So that's why you have this shift away from recommendations of so-many milligrams of cholesterol and more about just general healthy eating. Less junk food, more fruits and vegetables and stuff like that."

"But does the cholesterol I eat matter or not?" Jim asked.

"I mean it matters a bit, but like I said before, it's the cholesterol in your blood that matters more. Some people with high cholesterol change their diet and their cholesterol levels fall quite a bit, but in other people it barely budges."

"So I should go get my cholesterol checked?" Jim asked.

"Among other things, yes. Cholesterol isn't the only thing that affects cardiovascular disease. High blood pressure, diabetes, smoking, and how much you exercise all matter too. It's one part of the equation. We shouldn't overstate its importance but we shouldn't understate it either."

The waiter arrived with our bill, which he placed discreetly on the table. "I'm sorry I forgot to ask you all beforehand," he asked. "Do you want me to split the bill up or will you pay it all together?"

Jim was reaching for his wallet when the complexity of the problem dawned on him. He likely planned to pay for Katie's breakfast but what was he supposed to do about me? Would it be more awkward to offer to pay only for Katie's breakfast or to tell everyone to cover their own plate, frugality taken to an extreme? Katie, blissfully unaware of Jim's mental turmoil, folded her napkin and answered a message on her phone.

"Sir?" the waiter prompted.

"Charge it to my room if that's okay," I suggested.

The waiter directed me to fill out the slip of paper and asked to see my room key. We gathered up our belongings and stood to leave.

"I'm just going to go to the washroom," Katie said. "I'll leave my stuff here."

She walked off, and Jim turned to me. "Thanks for picking up the tab. I just froze back there. I didn't know if it was presumptuous for me to offer to pay for breakfast or rude if I didn't. I never know what the right thing is anymore."

"I don't think anyone does. There's no rule book. Just be yourself and be kind and you won't go wrong. You rarely go wrong by being a nice person."

"Well thanks again. You are the best wingman ever." He punched me in the shoulder. "I'll e-transfer you the money later today."

"Don't worry about it. It's fine. But after all the effort I put into this, you better marry this woman."

"Okay, I'll try."

"I was kidding."

"Oh right, of course. I knew that." It was an unconvincing lie, and I realized that compared to Jim, puppy dogs were better at hiding their feelings.

We were silent for a moment.

"Do you think it will bother Katie if I keep eating eggs? Like not every day, but at least sometimes?"

I thought about it for a second. "Well, scientifically speaking, it's not going to make a difference if you don't do it to excess. And when it comes to Katie, if you do it right, it probably won't matter either. People tend to forgive the small stuff when the big stuff is all in proper alignment."

Jim thought about that one for a long time while we waited for Katie to come back from the bathroom.

MYTH #8

Caffeine Can Trigger Heart Attacks

It might seem hard to believe that anyone could forget that they were supposed to be on stage giving a talk to a few hundred people, but that's what happened to me. As Jim and I stood in the hotel lobby waiting for Katie, I received a text from Alexi.

"Where are you?" it said.

I texted him back. "Lobby."

A minute later he came running up to me.

"What?" I asked.

"Do you know where you're supposed to be right now?"

I didn't, but the realization dawned. I swore under my breath and checked the time on my phone. "We still have five minutes. It's okay. These things never start on time."

"They never start on time because the speakers are always late. A self-fulfilling prophecy doesn't make things okay."

"Fair enough. Let's go." I turned to Jim to wish him well and we shook hands.

"There's a chance I might be in your neck of the woods next month . . . might fly up to see Katie," he said.

I told him to say bye to Katie for me, but she came up behind us.

"Katie said you were the most interesting person she ever sat next to on a plane," Jim said.

Did he not realize the comparison also included him?

"I did," Katie said. "Although the bar is pretty low. You kept your shoes and socks on for the entire trip and didn't try to physically assault the flight attendants. That counts for a lot these days."

We had another round of goodbyes and then Katie and Jim headed for the valet stand so Jim could collect his car. Alexi and I double-timed it to the conference center.

"Were those the two people from the plane?" Alexi asked.

"Yes, Jim and Katie."

"She's gorgeous."

"You're engaged."

"I mean you should have asked her out. She's beautiful, and she lives in the same city as you."

"Is that what you base relationships on these days?"

"It's a good enough place to start."

"I think Jim is smitten already."

"Let me give you some advice."

"Please don't."

"If you find a beautiful woman who is willing to listen to you drone on endlessly about statistics and medical research —"

"Hey!"

"And actually tolerates your company for extended periods of time, then you need to ask her out."

"Duly noted." We got to the conference room. An attendant at the door scanned our lanyards to let us in. Alexi grabbed a seat and I headed to the front of the room to give my presentation. After I was done, I found him in the same seat, and we skipped out before the next talk.

"It was good," Alexi said.

"Yeah?"

"Uh huh. Nobody threw any tomatoes at the stage."

"They probably didn't have any tomatoes."

"Probably. You had breakfast with . . . ?"

"Jim and Katie. Yes. I take it you haven't eaten?"

"No. Just got up before your talk, and I have a meeting soon."

I remembered that he'd mentioned it last night. "We can grab something from the coffee shop next door if you're in a hurry."

"Yes. Good. That way I can meet your barista."

"She's not my barista, and she probably won't be working today anyway."

But as it turned out, she was working. The shop was empty, and Casey sat on a stool behind the counter reading something on her phone. She looked up when we entered and smiled. We walked up to the counter and sat side-by-side. Casey greeted me enthusiastically. Say what you will about independent coffee shops, you just don't get personalized greetings like this in the big chains. Then she looked at Alexi.

"Who's this?" she asked.

I told her his name and explained we went to school together. She told Alexi how happy she was to meet him.

"What happened with your other friends? The ones you met on the plane?" she asked.

"Oh, Jim and Katie. I just saw them."

"Did they hook up?"

I was taken aback. "I . . . I don't really know to be honest —"

"They totally hooked up," Alexi said.

"You don't know that for sure," I said.

"I'm inferring from what I saw."

"Is Katie glamorous and fancy?" Casey asked. "In my mind, she's glamorous and fancy."

"Yeah," Alexi said. "She is."

Casey frowned and looked at me. "You should have asked her out."

"That's what I said!" Alexi said. He'd found an ally.

"I think Jim is head over heels in love with her," I said.

Casey giggled. "In that case, I hope they get married and then they'll invite you to the wedding and you can give a speech at the reception about how they met. And then everyone there will lose their minds and buy you drinks all night."

"You really have this all planned out in your head, don't you?"

"Yep!"

"They only met yesterday," I said. "We might be putting the cart before the horse."

"When it's the right person, you know. You know right away. You may not be able to admit it to yourself at the time, but you know. You think about them when they're not around, you get a knot in your stomach when you think you might see them. There's something. You know when it's right."

A few minutes ago, I could have thought up a number of counter-arguments, but in that moment, they struck me as unconvincing.

"Don't worry," Alexi said. "I'm working on him." He ordered a breakfast sandwich and a large coffee.

"Let me guess. Black?" she asked, suppressing a smile.

"Um, sure. Okay." Alexi eyed the breakfast sandwiches in the display. "Are those sandwiches fresh? I mean were they made today?"

"My manager said I'm supposed to answer yes if anyone asks. So, yes."

Alexi considered his options. "I'm not getting up again. Give me the one with egg and bacon."

"Is that healthy?" Casey asked.

I was primed to give an encore performance of breakfast, but Alexi beat me to the punch. "I don't really care."

"That's the spirit!" Casey said. "Eat what you want to eat. You only live once."

"That's my philosophy," Alexi said.

"It's not the healthiest breakfast," I said.

"First you come for my wine, and now it's my eggs. What's next? Coffee?"

"No,, we covered that yesterday," Casey assured him. "We had a long talk about this."

"About coffee?" Alexi asked.

"Yeah, we talked about coffee and dating apps and admission to college and it was a bit all over the place, but it all tied together in the end. And I learned a new thing. Selection bias. That's why online dating is a soul-crushing experience. It's not my fault in any way."

"So to be clear, you listened to him drone on for hours about medical research and statistics?"

"Oh God yeah! It was fascinating. We talked about coffee and research and the WHO and then we did an experiment and measured the temperature of a cup of coffee. I haven't had this much fun at work since the day I got hired and I realized I'd be able to pay my rent again."

"So do you drink coffee with a clear conscience now?" I asked her.

"Not really. It makes my heart all fluttery. I don't think caffeine is good for my heart. Actually, is that true? Is that a real thing or is that just something people say?"

"Don't encourage him," Alexi said. "He might go off on this topic for a full hour if we don't keep him in check."

Casey giggled and reached below the counter to get one of the sandwiches. She grabbed a mug from behind her and poured Alexi his coffee. She looked at me. "I'm going to give you one too. You look tired." Given the combination of sleep deprivation and jet lag, I wasn't terribly surprised.

"If you keep plying him with coffee, he'll be fine." Alexi took a bite of his sandwich.

Based on his expression, it was neither as good as he hoped nor as bad as he feared.

"I just remember hearing about that teenager who died of a caffeine overdose. It's hard to hear stories like that and not think that maybe we would be better off not drinking coffee," Casey said.

"I don't remember hearing that," Alexi said with a mouthful of sandwich.

"It happened," I told him. "Two teenagers died. Both cases were later found to be caffeine overdoses. In one case the teenager drank several different caffeinated drinks and the other was using a caffeine powder supplement."

"So these were accidents?"

"Well there was no suggestion it was deliberate, if that's what you mean. In both cases, the teens just didn't realize that excessive caffeine consumption can be dangerous."

"Given how many people drink coffee on a daily basis, I'm skeptical that caffeine can be dangerous. If it were that bad for you, humanity would be finished," Alexi said.

"I lay off the stuff regardless," Casey said. "Not worth the risk, and I don't really like the taste."

"Is there a risk though?" Alexi asked. "Those seem like isolated cases, and I don't remember anybody in medical school saying that caffeine is bad for you."

"Caffeine *is* a stimulant," Casey pointed out. "It can't be good for your heart. Because my friend Cindy started taking ADHD medication. She was actually getting it from some guy at university who was selling his prescriptions. Apparently, a lot of people would do that during exams. It was super sketchy. But anyway, she eventually stopped because it was giving her palpitations, and her doctor said that it was probably the pills because any stimulant can make your heart beat faster."

She was right. Caffeine is a stimulant, and any stimulant can make the heart beat harder and faster, at least to some degree. But there's a difference between a slightly faster-than-average heart rate and actual cardiac damage. Proving cardiac damage was more difficult.

"Has anybody ever checked to see if drinking coffee damages the heart or whatever?" Casey asked.

Could she read my thoughts? "There have probably been hundreds of studies — I'm tempted to say thousands of studies — about coffee over the centuries."

"Centuries?" Casey asked.

"If you wanted to, you could go all the way back to the coffee experiment of King Gustav III of Sweden in the 18th century."

Casey and Alexi looked at each other. "Don't do it. It's a trap," Alexi said.

"It's a fascinating story," I told her.

"It really isn't," Alexi said. "It never is."

Casey scrunched up her face and weighed her options. She took a deep breath. "I want to hear about the Swedish king and the coffee experiment. I have a feeling this is going to be bonkers. It's going to be bonkers, isn't it?"

"A bit," I said. "You see, Gustav III wanted to prove that coffee was dangerous so he designed an experiment. He got a pair of identical twins and ordered that one twin would drink coffee every day for the rest of his life and the other would drink tea. Whichever one died first would prove the point."

"And why would these twins agree to this?"

"They were both prisoners who were going to be executed. It was either agree to the study or be killed."

"Ah. I see. In their place, I'd have done the same, given that the alternative was being killed. So who won?"

"The tea drinker died first but many years later, in his eighties. The coffee twin died sometime later."

"The king must have been none too happy."

"I think he was dead at this point. He was assassinated during a masquerade ball."

"I guess if you're going to assassinate a king, you want to do it in a setting where everybody is wearing a mask so you can get away easily."

"He actually survived the bullet shot and only died of infection a week later. He had enough time to round up all the nobles who were plotting against him."

Casey contemplated the story and looked at Alexi.

"I warned you," he said.

"I stand by my decision to ask about the king of Switzerland," she said.

"Sweden," I said.

"Whatever. But I have some questions. Why did they keep doing the study even after the king died? And come to think of it, isn't it a bit suspicious that you had two identical twins in prison, at the same time, both going to be executed? That seems a little too convenient."

"This whole story might be apocryphal. But the point is, people have been talking and debating and studying the relative merits and dangers of coffee for centuries."

"But for people to keep talking about it, over and over, some of the research must say that coffee is bad for you. Otherwise, people would have moved on to other things."

"You're right. There have been some studies suggesting that coffee is dangerous, and others suggesting it's not, and others suggest that it might even be good for you. You really have to interpret the numbers in the right way."

"Oooooh! Are there going to be more tables and graphs like last time?" Casey asked. "Wait. I'll get the napkins."

Alexi shook his head. "Don't encourage him."

Casey ran off to the back room and came back with a stack of napkins and pen. She slapped them down on the counter. Alexi folded his arms and shook his head.

Facing no overt objections, I proceeded. "Let's say you want to see if coffee can trigger a heart attack. You take a bunch of people who had heart attacks and see which of them were drinking coffee and when."

"And you compare them to people who didn't have heart attacks, right?" Casey asked.

"In our scenario, you can compare them to themselves. It's a statistical technique where everybody can become their own control group."

"That's kosher?"

"It only works if the thing you're looking at is something that can happen repeatedly, like a car accident. You can get into a car accident now and then again a month later. It wouldn't work for things like death because you can't die twice."

"Wasn't that a James Bond movie?" Alexi asked.

"You're thinking of *You Only Live Twice*," Casey said.

Given that Alexi and Casey were talking Bond, I wasn't sure I was doing a good job explaining the case-crossover study design. It doesn't work for everything, but it can be extremely useful in specific situations.

It was designed to study risks that come on quickly and disappear quickly without any lingering effects. The classic example was studying whether using your cell phone while driving causes accidents. It's a perfect example because your phone is dangerous when you're using it and harmless the minute you put it away.

For that reason, coffee would also be a good candidate for a case-crossover study. The risk of having a heart attack, if there was a risk, would go up when you started drinking the coffee and would disappear after an hour or two when the caffeine has been metabolized by your system. At which point, you're back to where you started.

Eventually, Casey and Alexi stopped talking about their favorite movies. Casey asked me to tell her about the new study design I'd mentioned. When I finished, she looked confused.

"But what about all the other things that can affect your risk of having a heart attack, like smoking, diabetes, exercise, and all that?"

"That's the beauty of this type of study design," I said. "You're comparing yourself to yourself so all those things cancel out. Whatever was true before you drank the coffee is still true after you drink the coffee."

"Clever."

"It doesn't work for every situation, and if you don't use it the right way you can really mess things up."

"Kind of like a nail gun," Casey said.

"I . . . suppose so." I had to admit that it was an apt analogy. "So with all those caveats in place, researchers have studied coffee this way. They've looked at the risk that you'd have a heart attack in the one hour after you drank a cup of coffee."

"And what happened?"

"They found that the risk of heart attack went up by 50 percent."

Casey's eyes widened so much, she reminded me of an emoji. "You realize I only drank coffee last time because you told me it was okay," she said.

"He has a habit of leading young women astray," Alexi said between bites.

Casey pouted. "It's always the nice ones."

"You don't seem any worse off for the experience," I said.

"Yeah, but it was a wild afternoon. I had tons of nervous energy. I cleaned everything around here and got so much done. I imagine that's what being on cocaine feels like. I was almost bouncing off the walls."

"I'm glad you came back to Earth in one piece."

"Does the risk of a heart attack wear off quickly too?"

"It did in the study. The heart attack risk increased by 50 percent in the first hour but there was no increased risk by hour two or hour three."

"Then I'll monitor you two for the next hour." She giggled. "Just to make sure."

I thought she was joking, but she actually set the timer on her phone. "It's not necessary," I told her. "That 50 percent number can be a bit misleading."

"Numbers usually only mean one thing. Fifty means fifty."

"But when you say that something has increased by 50 percent, you have to ask yourself 50 percent *from what*?"

"I'm not following you." Casey leaned her elbows on the counter.

"Let's accept the basic premise that a cup of coffee increases your risk of heart attack by 50 percent. The question you have to ask yourself is, What was your risk before you drank that coffee?"

"You're going to have to tell me," Casey said. "You're the doctor."

"So we can calculate your risk of having a heart attack. There are formulas."

"I was told there would be no math."

"There's always math with this guy," Alexi said.

He finished the last of his sandwich. He seemed unsatisfied, so Casey asked him if he wanted another one. He hesitated and grunted, which in coffee shops seems to imply tacit agreement, so she brought him another breakfast egg sandwich and refilled our mugs.

"Remember I was an English lit major. I don't do math," Casey said.

"You're an English major?" Alexi asked.

"I wrote my thesis on Chaucer and the differences between the West Midland and London dialects of Middle English."

"That's really fascinating."

"Best decision of my life. Those four years in grad school really prepared me for my career in the food service industry." Casey looked around the coffee shop and seemed to contemplate her existence.

"Is there anything we can do?" Alexi asked.

"Do either of you have a time machine so I can go back and tell 24-year-old me to go to law school like my parents told me to?"

Neither of us did.

"Well, it doesn't matter I guess. Things tend to work themselves out in the long term. Plus you always meet interesting people. I mean one day you're just sitting here minding your own business, wondering what series of life choices got you to where you are now, when a jet-lagged doctor walks into your life and tells you that coffee doesn't cause cancer and that the reason why everybody on dating apps is horrible is because of statistics or selection bias or whatever and not because men just generally suck and then you realize that the world isn't such a bad place after all and maybe things are going to be okay."

There was a compliment in there somewhere. Alexi smiled and tried to hide his grin behind his coffee mug.

Casey wiped down the counter. "Just no math."

"I'll keep it simple, I promise."

"I'm going to need a pinky swear on that."

I didn't know what she meant but apparently interlocking our little fingers made my promise more concrete and binding to the satisfaction of all involved, especially Alexi, who seemed thoroughly amused.

"I don't think pinky swears hold up in court," Alexi said.

"Well I didn't go to law school in the end, did I?"

"Touché," he said.

"Anyway, coffee, 50 percent increase in heart attacks, but probably not. Go." Casey clapped her hands twice in quick succession. Alexi snorted coffee out of his nose.

"Right, so the problem with this study was the issue of relative risk vs. absolute risk. I don't know if you're familiar with this."

"Why in God's name would I be familiar with this? I spent my college time reading *Beowulf.* Do you want to have a debate about what language *Beowulf* is written in? Because then I'm your girl. And let me tell you, it sure as hell isn't English. You think Shakespeare is hard to read? Dude, you're going to think this thing was written on another planet."

"Let me put it to you this way. Let's say I told you that I had a system that made it twice as likely that you'd win the lottery."

"I'm not going to lie. That would really solve a lot of my short-, medium-, and long-term problems."

"But think about that offer mathematically."

Casey pursed her lips in concentration. "No."

"I mean think of it this way. Your chances of winning the lottery if you have to pick six numbers out of a possible 49 are about . . . 1 in 13 million 980-something thousand."

"I like how you said 'about' before that."

"Let's say one in 14 million. If you double your chances, it's still only two chances in 14 million."

"So that's why buying bunches of lottery tickets doesn't work."

"No."

"I guess I need an alternative retirement plan then."

"The problem is your chances of winning the lottery are pretty low. Even if you double them, they stay pretty low. And the same thing happens in medicine. A rare outcome stays rare even if you double or triple or quadruple the risk."

"I get you, but if you told me that something tripled my risk of getting cancer, I'd be worried about getting cancer."

"That's a normal reaction. But if you look at the numbers, sometimes scary things don't look that scary."

"I think Casey warned you about the math," Alexi said.

"That's true. We pinky swore."

"I remember. But I can keep this simple."

Casey looked at me skeptically but waved her hand for me to continue.

"There was a study a while back looking at whether strawberries could help prevent heart disease in women."

"That's my kind of study," Casey said.

"That's understandable. Telling people tasty things are healthy is an easy sell. So when the study showed a 32 percent reduction in heart attacks, the headlines wrote themselves. 'Berry Good for Your Heart' and 'A Tasty Way to Prevent Heart Disease.' Stuff like that."

"Very clickbaity. And I guess you're going to tell me that the 32 percent number was wrong."

"Let's assume it was right. Let's pretend there weren't any problems with accurately measuring how many strawberries the women ate or that people who eat lots of fruit are probably healthier than people who don't eat fruit in other ways."

"Pretending problems don't exist is how I manage to get out of bed every morning. It's the only practical way to deal with student debt."

"Now, let's just take that 32 percent number at face value. It sounds impressive but you can also look at it a different way."

"I'm all ears," Casey said. She plopped her elbows down on the counter and rested her head in her hands right in front of mine.

She was close enough I caught a whiff of what I assumed was her shampoo. I thought it had a faint fruity smell and I temporarily lost my train of thought. "Ummm . . . yes. Well the thing about that number is . . ."

"Cat got your tongue?" Alexi asked.

"No, I'm fine," I said. I cleared my throat. "Consider this. Saying that you're 32 percent less likely to have a heart attack sounds impressive, but how likely were these women to have a heart attack to begin with?"

"I'm assuming that's a rhetorical question," Casey said.

"It's quite low because this study was done in women between the ages of 25 and 42. That's a group of women who are at pretty low risk of heart disease."

"That's possibly the most reassuring thing you've said to me all day. I knew there was a reason I kept you around."

I knew she was joking, but for some reason, I still enjoyed hearing her say that. "This was a study of almost 100,000 women and they were followed for 18 years. And in all that time there were only 405 heart attacks."

"That doesn't seem like a lot."

"It's less than a tenth of one percent per year when you average it out. So when you say the risk of having a heart attack was 32 percent lower in the group that ate more strawberries, the difference really amounts to a few hundredths of a percentage point in absolute terms."

"It's hard to wrap my head around what a fraction of a percentage point actually means."

"There are better ways to ask the same question. You could ask, 'How many women do I need to convince to start eating strawberries on a regular basis in order to prevent one heart attack?'"

"Convincing even one woman to start eating strawberries is a win in my book."

"With whipped cream," Alexi said.

"Yes! Good idea!" Casey said.

"You're not helping," I told him.

"I never said I was trying to be helpful."

"I think it's worth reminding everyone here that strawberries have a lot of sugar in them and that coupling them with whipped cream isn't a healthy eating choice," I said.

Both Alexi and Casey booed in unison, heckling me.

"The point is that when you do the math" — I held up a hand to forestall another round of booing — "you basically have to get around 5,000 women and feed them strawberries every day over the course of a year to prevent one non-fatal heart attack."

"I wish someone would feed me strawberries," Casey said. She looked out the window with a dreamy expression. "Or grapes. That would work too. To be honest, the type of fruit is irrelevant."

"Point is that a 32 percent decrease in heart attacks sounds impressive. But saying that thousands of women need to start eating strawberries to prevent just one cardiac event is a lot less impressive."

"Yeah, I suppose," Casey said. "Shame though."

Alexi had finished his sandwich, and she took his plate and disappeared it behind the counter. It was an automatic, almost subconscious gesture and she did it without breaking the stride of the conversation.

"So basically you can measure how risky something is in different ways?" she asked.

"Either relative risk or absolute risk. The relative risk tells you whether something is twice as bad or half as bad as something. The absolute risk tells you how likely something is to happen to you."

"I kind of feel like the second one is the more relevant of the two because, honestly, that's what I want to know."

"Most people would agree with you," Alexi said. "The absolute risk is more concrete. It's easy for people to understand. If 5,000 women eat strawberries, one heart attack is prevented. That makes sense to people."

"So why use relative risks at all?" Casey asked.

"Two reasons," I said. "First, mathematically relative risks are much easier to work with. When you do analyses, that's the answer that comes out of the equations." Casey yawned. It was probably a coincidence, but it still felt like a veiled criticism. "The other reason is absolute risks are dependent on the population you study. In high-risk populations, you see greater risk reductions with treatments. In low-risk populations, it's not as big."

"So medications are more effective the sicker you are?" Casey asked. "That seems counterintuitive."

It does seem counterintuitive at first, but I walked Casey through an example to prove the point. Alexi knew all this, so he listened quietly. But the concept makes sense if you consider antibiotics. If you had sepsis, a severe blood infection, then antibiotics would probably save your life because you'd certainly die without them. But if all you had was mild bronchitis, and you were otherwise healthy, you'd probably pull through regardless. The antibiotics would speed your recovery and likely keep the bronchitis from turning into pneumonia, but you probably wouldn't die. Antibiotics do the most good the sicker you are and the more severe your infection. In a low-risk patient, the magnitude of the benefit is much less.

"So the absolute risk isn't constant for everybody?" Casey asked.

"No, and that's the second reason why relative risk reduction has some value. It's largely independent of the baseline risk. That's not completely true, reality is more complicated —"

"Reality generally is," Casey said.

"Fair. But essentially, that's the point. Relative risk survives and keeps being used because it's mathematically convenient and, theoretically at least, it's superimposable on different groups of patients."

"Bit misleading though. Telling me that eating strawberries cuts heart attacks by 32 percent is really impressive, but seeing it's only a 0.00000-whatever percentage point decrease is really . . . blah. If I were a cynical person, and I am because I got a PhD in English literature and I'm working in a coffee shop, then I'd think people use relative risks because it sounds really impressive when the reality isn't."

"There probably is some of that," Alexi said. "The truth is even many doctors and scientists don't know the difference because they don't have a stats background. I just treat patients all day, I don't get involved with this stuff. I know about it only because this guy won't shut up about it." He pointed his thumb at me again.

I came up with a really good comeback, but by the time I thought of it Alexi had moved on.

"The reality is that most people don't have the time or inclination to start dissecting papers to tease out all these subtle statistical points. They'll see a 32 percent reduction and take it at face value," he said.

"Sounds like you need to stay close to this guy," Casey said. "Otherwise, you'll go around telling women to eat strawberries. Which wouldn't be a bad thing necessarily but maybe not that medically justifiable."

"It's really the only reason we're friends," Alexi said. He looked sideways to see if I'd rise to the bait. "That and sometimes there's a news story about something that causes cancer and my girlfriend freaks out and needs someone to tell her that it's not true."

"Fiancée," I said quietly.

"Damn it, I keep doing that. I just got engaged and I haven't gotten used to saying *fiancée* yet."

Casey clapped her hand over her mouth and stared Alexi down. "I'd fix that reeeeeeeeaaaaaaaaaaaally quickly," she said. "Women put up with a lot in relationships. Trust me. I'm on a first-name basis with the guy who works at Ben & Jerry's around the corner for exactly that reason. If you keep referring to your *wife* as your *girlfriend*, well let's just say there's a limited number of times that can happen before things start going badly for you."

"I know. I know. I'm working on it."

"What kind of stuff freaks out your girlfriend — dang now I'm doing it. What kind of stuff freaks out your fiancée?"

"So there was this study a little while back about how using hair straighteners could increase your risk of endometrial cancer —"

Casey let out a small wail and dropped her head to the countertop. "As if having a uterus wasn't bad enough. Now I'm going to get cancer if I straighten my hair when I get dressed up to go to a wedding."

I thought I should say something reassuring to her and I figured I'd start with some mild self-affirming positivity. I wanted to tell her that I thought she looked good with curly hair and that she didn't really need to straighten it. The gentle curls bounced softly when she laughed or turned her head to the side and it made me think of —

"It actually wasn't that bad," Alexi said. He looked at me.

He expected me to follow through on the lay-up. "Yes, it wasn't that bad. It was the same sort of problem as with the strawberries. The study found that women who used hair straighteners had an 86 percent increase in uterine cancer. Now that sounds scary, but that was the relative risk. In absolute terms, the risk rose from about 1 percent to a little bit less than 2 percent. That's only a one percentage point change."

"Yeah, but 1 percent isn't 0 percent."

"No, it's not. But there are some caveats."

"You like saying the word *caveats*," Casey said.

"Well, I mean it comes up a lot when you have to dissect news headlines about health because there's so much . . . ummm . . . nuance —"

"This is our new drinking game. When he says *caveats*, we take a drink," Casey told Alexi.

"You keep alcohol at work?" Alexi asked.

"Oh. Right. No. They were pretty clear about that at the meeting."

"The point I was trying to make is even when you put the risk into an appropriate frame of reference, there are still many other caveats — damn it."

Casey laughed.

She was getting inside my head. I needed to look up some synonyms later.

Casey took pity on me. "Okay, tell me what you were going to say."

"The risk was small to begin with. And it wasn't seen in women who used hair straighteners only occasionally."

"Define *occasionally*," Casey said.

"Women who used them less than four times per year."

Casey made a face, but I couldn't tell if she thought four times per year was a lot or not. Alexi was no help. I picked my next words carefully to avoid saying *caveats*; naturally my brain was trying to put it into every sentence. "The other . . . issue is they studied a bunch of different hair products and only saw a possible cancer risk with hair straighteners. There was nothing for perms, bleach, or hair dyes. With highlights, the risk of uterine cancer was slightly lower."

"Finally! Some good news. I've been meaning to get highlights for a while now. What do you think of blond?" she asked. She spun her head side to side so that her hair bobbed back and forth a bit and she twirled to show us the back of her head.

I didn't think that she would look good with highlights. Her reddish-brown hair offset the brownish hazel tint of her eyes. On top of that, the highlights would probably —

"The lower risk with highlights didn't really hold up during the statistical analysis though, did it?" Alexi asked.

Right. We were talking about cancer risk. "No, it didn't," I told them. "I don't think you can put much weight on it. They looked at a bunch of different hair products and most had no effect on cancer risk. One had a slightly lower risk, one had a slightly higher risk. Just by random chance, if you do enough tests and analyses, you'll find some positive statistical variations."

"So even if the risk is real, which it may not be, it's pretty small," Casey said.

"Exactly."

"So if you want to straighten your hair because you're going to your 15-year high school reunion and you want to show that jerk Tim Bobsworth that you're over him and he's missing out by having been such a self-absorbed narcissist, then it's fine."

"That's a very specific example," I said. "But yes, that's it. Also worth pointing out that you don't need to straighten your hair to make this guy jealous. You can just be yourself."

Casey smiled, but it wasn't accompanied by her usual giggle. It was one of those deep smiles that spread to her eyes. "And the reason you told me this story is because coffee is the same thing, right? The study says there's a 50 percent increase in heart attack risk, but really it's not that high."

"No."

"The 50 percent increase is a relative risk increase. It's half as high again. But how much does it really go up in the absolute sense?" she asked.

"That's the question. In fact, when the study came out somebody wrote an accompanying editorial on exactly this point. It was really funny."

"Funny like ha ha funny, or funny like only-to-other-scientists funny?"

Admittedly, it was probably the second one. "It was funny because he wrote the editorial like a letter to the editor you might see in the newspaper. He started off by saying he was having breakfast and was about to drink his morning coffee when he saw the study and wanted to calculate how much risk that one cup of coffee might mean to him. So he went through all the math of calculating cardiac risk —"

"We're skipping that part," Casey said.

It was a statement, not a request. "Fine. But the important part is that we calculate cardiac risk in terms of ten-year increments. High risk is 20 percent over 10 years. If you want the risk of a heart attack in the 60 minutes after drinking coffee, you need to take the average cardiac risk and spread it out over 20 years times 365 days per year times 24 hours per day."

"I guess you're building toward a really small number here."

"I am. It works out to a 0.00005 percent increase in the one hour after a cup of coffee."

"This doesn't mean anything to me. It's such a small number that my brain can't process it."

"We can do what we did with strawberries. We convert it into the number needed to treat, or in this case the number needed to harm. We flip it around. Instead of asking how much does one cup increase your risk of heart attack, we ask how many cups of coffee do you need to drink to cause one heart attack."

"That's more useful."

"When you work out the math, it means one extra heart attack in the one hour after drinking 2 million cups of coffee. He capped off the editorial by saying that by the time he finished doing all his calculations, his coffee had gone cold." I smiled and paused for effect at the punchline.

"Womp, womp." Casey said. When she saw my puzzlement, she explained. "That's a sad trombone."

"She's mocking you," Alexi said.

"I am," Casey said. "But only because you think statistics can be funny."

I wanted to explain to them that statistics was a deadly serious business, but I worried that might be counterproductive.

"Worth remembering too that there's a lot of research showing that coffee isn't harmful and that normal amounts are protective," Alexi said.

"So coffee is actually good for you?" Casey asked. "That surprises me, since it makes me all fluttery and unsettled."

"A lot of it is observational research," I said.

Casey nodded, murmured non-specific agreement and mumbled things like "Ah yes" and "observational" and "of course" while intermittently clucking her tongue and affecting a British accent in later iterations of the cycle.

"She's mocking you for your use of jargon," Alexi said.

"Observational research is when you don't actually do anything to the people you're studying," I said. "You just sit back and observe what they do. When you do that, the people who drink coffee often have a slightly lower risk of heart disease than the people who drink zero."

Casey asked a few follow-up questions about the differences between observational research and randomized trials, but the names themselves sort of give away the definitions.

"Except, what about the people like me who don't react well to coffee?" she asked. "They probably avoid it and don't drink any."

I agreed.

"So all these people with heart issues and arrythmias drink no coffee and get lumped into the zero coffee group."

It was a remarkably astute observation and both Alexi and I said so.

"We were discussing that last night, actually," Alexi said. "Same thing happens with red wine. It looks like red wine is good for you, but that's at least partially because people who have medical problems are usually advised to avoid it, so the people who abstain from alcohol are at higher risk to begin with."

"Makes sense. Now, I imagine we'll be calling this phenomenon the Casey Effect moving forward?"

"I don't think I have the authority to rename statistical concepts," I said.

"You won't know until you try."

"I'll see what I can do. But whether we call it the Casey Effect or reverse causation, the point is the same. Coffee looks beneficial but it might be because people with heart problems are more likely to avoid it. In any case, it's probably not harmful in reasonable doses."

"A reasonable dose being less than two million cups?"

"Less. Even by liberal estimates, at ten cups per day, you're starting to get into dangerous territory."

"Goodness! Do people actually drink ten cups of coffee per day?!?" Casey said. "How can a human being tolerate that much caffeine?"

"We both went to school with people who got close to that number on a daily basis," Alexi said.

"It's true," I said. "Some people lived on the stuff."

Casey was having none of it. "I don't think that's healthy. Weren't their hearts going berserk? Couldn't that cause arrhythmias or palpitations or whatever?"

"It's a little unclear whether coffee actually increases the risk of serious arrhythmias," I said.

"Ahem." Casey didn't clear her throat. She actually said the word *ahem.* "I offer myself as exhibit A. Coffee does give me palpitations."

"A lot of people are sensitive to coffee and caffeine. And if it bothers you, you should avoid it. But it may not cause severe arrhythmias."

"Pretty sure it does," Casey said.

I did know of one study that tried to settle the issue. The CRAVE study had people wear monitors so they could record their heart rhythms rather than just relying on people to report their symptoms, the advantage being that continuous heart monitors will pick up everything and are the only objective way to look at electrical disturbances of the heart.

The researchers told patients they could drink as much coffee as they wanted to on one day but zero coffee on the next day. They alternated the pattern so they had a one on, one off system. Casey pointed out there was no way to be sure the people followed this protocol. They might drink coffee on days they weren't supposed to, which would invalidate the results. I admitted that was a problem, but the researchers did the best they could to control for this possibility.

The researchers sent participants daily reminder texts to let them know if it was a coffee day or a no-coffee day. They asked people daily about their coffee consumption. People had to push a button on the heart monitor every time they drank coffee so the researchers could link the coffee back to any arrhythmias. And everyone in the study had their coffee purchases reimbursed if they submitted receipts. They used the GPS on people's phones to see when they went to coffee shops.

Casey considered everything. "Those are some intense Big Brother–level monitoring measures. I'm not sure how I feel about that. That could get annoying after a while and I'd probably start tuning them out."

"It was only for two weeks. And I think the researchers probably did as much as they could to make sure people stuck to the one-on-one-off schedule."

"Fine, we'll give them a pass." Casey giggled at her own joke. I had to admit, I liked her laugh, and I couldn't help but smile too. It took me a second to realize Casey was waiting for me to say something.

"So what did the study show?" she asked.

"The study?"

"Yes, the study we've been talking about for the past five minutes."

Alexi laughed and clapped me on the back. "You're doing fine, don't worry."

"There were no significant arrhythmias associated with coffee consumption," I said. "Nothing you'd necessarily treat."

Casey frowned. "Why would you not treat an arrhythmia? That sounds dumb."

"Not every arrhythmia *needs* to be treated. Some are totally benign. If you take random people off the street and put heart monitors on them, you'd always see skipped beats here and there. Minor stuff like that doesn't need to be treated."

Casey stared me down for about 15 seconds. "I'm going to accept this because *you're* the one saying it. But it seems sketchy to me."

"You're putting a lot of faith in someone you met yesterday," Alexi pointed out.

"Most doctors I see are just jerks. Sometimes patronizing jerks. But you two seem cool."

Alexi turned to me. "She might be the first person to ever call you cool."

Casey giggled and hid her smile with her hand. "Nah, you two are okay. Plus we were talking for like hours yesterday. So we're totally friends now. It's all good."

Alexi looked at me but said nothing.

Casey turned back to me. "You're sure coffee is fine."

"The people in the study had minor electrical abnormalities. Usual stuff you'd see in anyone. But there was no difference between the coffee and no-coffee days in terms of serious arrhythmias like atrial fibrillation."

"You can't just make up funny-sounding words," Casey said.

"Atrial fibrillation is a type of arrhythmia that can increase your risk of stroke. But it doesn't matter. It didn't happen. Coffee didn't cause that type of arrhythmia."

"So all good then?"

"Not exactly."

Casey slapped her hands down on the table. "Has anyone ever told you you're kind of a tease?"

"You'd be surprised," Alexi said.

In our youth, I probably would have boxed his ears.

"Aww." She cupped my hands in hers. "We're making him feel bad, we have to stop."

"I have a thick skin," I assured them.

She looked down at my hands and turned them over in hers. "I don't know, you have really soft hands for a man. I mean really soft. You've obviously never done manual labor in your life."

That was true. I figured I should say as much but at some point, I'd have to retrieve my hands and I wasn't sure how I was going to go about that. Casey kept talking and her tendency to gesticulate for emphasis meant that she let my hands go. I'd been holding my breath.

"So there are some arrhythmias when people drink coffee?" she asked.

"Yes . . . no . . . I mean . . . on the days when people drank coffee there were no major arrhythmias. But there was one difference. When people drank coffee there were more PVCs."

Casey looked puzzled. "The stuff they make pipes with?"

"No, sorry, no. PVC stands for premature ventricular contraction. Basically, a heartbeat that comes in before it's supposed to. When people say they feel a skipped beat, they usually mean a PVC."

"And those are dangerous?"

"Depends how common they are. Isolated and rare episodes? No. But if you have long runs of PVCs or if they become very frequent then . . . maybe."

"So coffee does actually increase these . . ."

"PVCs."

"So, coffee *does* increase these PVCs. And if you have a lot and you feel them and your heart is all fluttery after drinking coffee, then maybe cut back and you'll feel better. Right?"

I told her she was right. Over the years, I've had a few patients who said they felt better and had fewer skipped beats after they cut back on their coffee. It was anecdotal obviously, but it seems to make a difference for some people.

"I think that proves my point," Casey said.

"I suppose so. But for most people drinking coffee probably has no negative effect as long as you keep to a reasonable amount."

"And a reasonable amount is 10 cups per day?"

"That seems a bit excessive. I would've said under five cups per day, and that's the recommendation from the FDA by the way."

"So how do you reconcile that with the cases where those young kids died?"

There's been a few examples of teenagers dying suddenly after drinking caffeine or energy drinks. They make headlines, but cases are mercifully rare. It's always hard to be sure if caffeine played a role. But there are some examples that are fairly clear-cut. One case in the UK involved two students who were taking part in a research study to see if caffeine improves exercise capacity. They accidentally got 100 times the dose they were supposed to get. They got quite ill but fortunately survived. There was another case of an Ohio teenager who died back in 2014. He was using a caffeine powder a friend had bought online, presumably as a workout booster. He died just before graduation.

"That's heartbreaking," Casey said. She looked like she might cry. "Any death is tragic but when it happens to someone that young it's just so senseless."

"I think he just didn't realize how concentrated the powder was. One teaspoon had as much caffeine as 25 cups of coffee."

Casey swore and covered her mouth. "Sorry, that was uncalled for but good God that's dangerous."

It was a fatal dose, I explained. The coroner later found he had a toxic amount of caffeine in his bloodstream and that had led to a fatal arrhythmia and seizure. The only good thing to come of it was the sale of caffeine powders was banned in Ohio and the FDA sent out a series of warning letters about the products.

"They should just be banned outright," Casey said. "Young people dying isn't worth it."

"Just to play devil's advocate," Alexi said. "It's not the product itself that's dangerous, it's the dosing."

I explained that the issue is most people can't measure out a safe dose properly. A proper dose is supposed to be 1/64 to 1/16 of a teaspoon. You can't measure that out accurately unless you have specific measuring tools at home which most people obviously don't. The risk for accidental overdoses is just too high.

"Couldn't they dilute it out or something?" Casey asked. "That way it wouldn't be as dangerous."

"What do you mean?"

"Like mixing the caffeine powder in water, or something, so it's not as concentrated."

"So . . . coffee."

Casey narrowed her eyes. "Nobody likes a smartass."

"I keep trying to tell him that," Alexi said.

"The point is coffee isn't dangerous," I said. "But caffeine in super high doses might be. So if you chug six extra large energy drinks or load up on caffeine powders, there's a certain risk."

"I'll stick with tea for now," Casey said. "I'm probably more sensitive to it than most people."

"You can do whatever your heart desires."

"That's good, I like it when people tell me I can do whatever I want without consequence."

"That's not exactly what I said."

"I think that's exactly what you said," Alexi said.

Casey nodded. "I'm glad we're agreed."

"I have my meeting to get to, so I have to head back to the hotel," Alexi said. He checked the time on his phone. I said we'd settle up and head back.

"When you come back tomorrow, we can talk about vitamin D. My friend Julie told me that I'm supposed to be taking a vitamin D supplement. So you can explain that to me. I'll have a lot of questions prepared."

"I'm flying out early tomorrow morning," I told her. "Heading back home. The conference is over today."

"Oh," Casey said. "I guess . . . goodbye then."

"It was nice meeting you." I wanted to say something else. That it was genuinely a pleasure. That I thought she was funny and fun to talk to and I'd enjoyed spending time with her. But I didn't quite know how to put it into words without it sounding weird or presumptuous. I decided discretion was the better part of valor and said nothing. "I might come grab a coffee tomorrow morning before I head out for the airport, but I leave pretty early."

"I only work in the afternoon," Casey said. Her voice was flatter than it had been earlier.

"Oh, I see. Bad luck then. Well, um, I guess we better get going." I looked at Alexi.

Alexi said nothing. He sorted his wallet and receipt in his pocket as we walked out the door.

When we reached the hotel's main entrance, Alexi looked at me. "You are quite literally the stupidest man on the face of the planet."

MYTH #9

Vitamin D Is the Cure for Everything

My final night in town was a quiet one. Alexi had a working dinner. I ordered room service and turned in early.

Despite my all-nighter and the day's early start, I couldn't fall asleep. My mind kept bouncing around from topic to topic. I felt a vague restlessness that kept me up until at least 2 a.m., the last time I checked the clock. When my alarm went off, I hauled myself out of bed, got ready, and headed downstairs. Alexi was almost certainly still asleep, so I texted him goodbye with a promise to see him back home. Within a matter of weeks, months at most, he would be back permanently, and we could relive our glory days again.

Except it wouldn't really be the same. He would be married soon. And probably have a baby eventually. So things would inevitably be different, no matter how much I might want to pretend otherwise.

I was about to ask the hotel reception to call a taxi but decided on one final pit stop. I asked him to hold my bags and I walked over to the Golden Lion.

I don't know what I expected. I knew Casey wouldn't be there. She'd said as much. But I felt like I needed to check. She wasn't. The figure behind the counter wore the same apron but a different name tag.

Greg was not nearly as friendly as Casey. He didn't smile. He asked more than once what kind of coffee I wanted. The implication was, order something or get out. I asked for a black coffee to go. He processed the order without commentary or critique. As I waited to pay, I figured I had to ask then or I never would.

"Is Casey here by any chance?"

"Who?" He punched keys on a terminal.

"Casey. She works here. Ummm . . . she's this tall . . ." I gestured with my hand. "She has light brown hair, hazel eyes, she's always laughing with this . . . bubbly giggling type of laugh. Oh, she has an English lit degree and —"

"Buddy, I don't need a biography. There's two women who work here. One wears glasses. One doesn't. Is she the one who wears glasses?"

"No."

"Then she isn't here."

"Is the one who wears glasses here? Maybe she knows Casey."

"No."

"Then why did you ask me . . . Can I leave a message for Casey?"

"You some weirdo or something? You stalking her?"

I would have answered no, but that was what a stalker would say. "I said I would leave her a business card."

Greg eyed me skeptically.

"She needed the number for this doctor." I took my business card out of my pocket and showed him.

Greg didn't want to give in, but he blinked and held his hand out for the number. Probably wanted to get back to trolling people on Twitter. "Okay, pass me the card and I'll leave it on the board for her."

I thought I should write something on it. I could write "It was nice to meet you," but the sheer banality of the statement put me off, and the realization that if it was pinned to a communal bulletin board people might ask embarrassing questions. I considered four or five possibilities and hit on what I thought was pure gold. "For all urgent vitamin D–related inquiries and medical questions. Remote consultations available. Graphics and charts provided on request." I handed the card over. Greg had no shame in reading the message. He didn't react since it meant nothing to him.

I only hoped it meant something to Casey. I didn't know how she'd react. She'd probably be confused at first then, hopefully, pleased. Then . . . what? I wasn't sure. Maybe I'd find out one day, but I wasn't going to learn anything in the coffee shop. I walked back to the hotel and got in a cab.

This time, nobody spoke to me at the airport. The only person who asked me anything on the plane was the flight attendant who wanted to know if I wanted the beef or the vegetarian pasta. I thought of Jim and Katie and our talk about red meat and whether it really caused cancer or not. I figured I should probably follow my own advice, at least some of the time, and I ordered the pasta. I was tempted to ask the flight attendant if it came with a side of bacon, but I realized he wouldn't know why that was supposed to be funny.

I made it home just before rush hour. The days were getting shorter, and the sun was just going down. I turned on my lights and cranked the heat back up and surveyed my apartment. I didn't figure anything would happen to my place while I was gone, but it was always good to check. My Christmas cactus seemed to be doing well, as was my African violet. I had watered them before I left and they both seemed unaffected and unconcerned by my absence. Plants are like that.

After a shower, I decided to make myself a cup of tea to remedy the additive effects of caffeine, sleepless nights, travel stress, and cramped airplane seats. As I put the kettle on, I noticed I'd missed a text, from an unknown number:

> You shouldn't have left without saying goodbye.

I tapped out a reply:

> You only say goodbye when you aren't going to talk to someone again.

I set the phone down and poured the hot water. I put on some soft music and stretched out on the couch. I stared at the cityscape as I sipped. I thought I'd slip into bed, read a bit and try to avoid a repeat of yesterday's insomnia. My phone pinged:

As I composed and discarded several possible replies, in my head, my eyelids got heavy. I slept straight through till morning.

I shifted over to look at the sunlight streaming through the window and felt something hard between me and the cushion. My phone had slipped from my hand when I fell asleep. My half-composed message, which I never sent, was now an endless string of random keys, followed by messages from Casey:

So . . .

I was promised answers to vitamin D–related questions.

I would also like graphs please.

There were too few graphs and diagrams last time.

Cause I need to know if I should be taking a vitamin D supplement. They're on sale at the drugstore.

?

. . . and you're asleep. Boo.

I yawned and stretched like a cat. Given the time difference between us, she probably wouldn't see a reply for a while. I'd use the time to think of something clever. I pondered what I should say as I ate breakfast, dressed, and drove to work. On my walk into the office, I drafted and deleted possible messages that lost their luster after a reread. I was stressed out by the time I fumbled for my office keys.

"No hello?"

I looked up and dropped my phone. "What? Oh sorry. I just need to finish this." I was overthinking the text. I just had to pick something and send it. I picked up my phone and typed the first thing that came to mind.

Vitamin D is not the cure to everything. Only love heals all.

As soon as I hit send, I regretted what I wrote. I realized now why good writers always work with great editors. I groaned.

"What's wrong?"

I looked up and realized that it was Irene, my secretary. She had been talking to me. "I just wrote something dumb. This day is not starting out well."

She seemed worried. "Maybe just sit down. Was the trip okay? Did anything happen? Do you want me to get you anything? Tea?"

Irene was my self-appointed therapist. She straddled the thin line between attentiveness and smothering. She also believed that there were few things that couldn't be fixed with hot beverages and homemade cookies. To my own shame, I rarely disagreed.

Just a weird morning, I said. It took three more assurances before she accepted that I was, in fact, alright. I gave her my trip receipts, and she walked me through the more urgent messages. Patients needed prescriptions renewed, results had to be signed off on, and some follow-ups had to be arranged.

At my desk, I was slogging my way through the emails when my phone buzzed:

You have the soul of a poet ☺

If cringe-worthy poetry went over well, I might be able to get through this without major embarrassment. Except now I was in the rather uncomfortable position of trying to figure out how to respond to a text that wasn't actually a question. Mercifully, three little dots appeared.

So . . . should I take vitamin D supplements? I ask because they're on sale. Also because I want to be healthy but mainly because they're on sale.

Short answer: no. You don't need them.

What's the longer answer?

The longer answer is there's a lot of research that vitamin D supplements don't really do anything.

But I thought we're all vitamin D deficient because we don't get enough sunlight cause we all work indoors to pay off our stupid student loans.

Not actually true.

So I *don't* have to pay off my student loans?

No, you definitely have to pay those back.

I could declare bankruptcy.

You still have to pay them back, even if you declare bankruptcy.

Stupid government thinks they're smarter than me.

Which in this case, they seem to be.

I meant the vitamin D thing wasn't true.

Believe it or not, most people are not vitamin D deficient.

WHAT!?! Are you saying those random people on Instagram lied to me?

kind of.

oooooooooooooooooooo

If you can't trust brittanny39687 then who can you trust?

Your doctor probably.

Yeah, but I have to make an appointment to see her.

Plus she charges money, and trust me we do not have good health insurance at work.

And Instagram is free.

I mean except for the fact that these companies are mining my personal information and selling it to God knows who.

But that's a long-term problem.

I'm a short-term problem type of girl.

Cause everyone with student debt is a short-term problem type of person.

My phone was undergoing an apoplectic fit of buzzing and beeping. Irene popped her head in to see what was going on.

"Everything okay?" she asked.

"Yes, just messages," I told her. "Nothing to worry about."

She reminded me I had a department meeting at 11. I checked my watch and saw that the morning was getting away from me. I turned back to my phone and had to scroll back to pick up the thread of the conversation. Casey was a bit erratic, but talking with her was never boring. It kept me on my toes.

That makes things more complicated.

But no, you're probably not vitamin D deficient.

My break is over, but I want proof.

I have to go to a meeting soon too.

K. Text me tonight.

My phone lapsed into an eerie silence, as if exhausted. And I now had no excuse but to get back to work. The first day back from a trip is always a difficult one and it's easy to get mired down in minutiae. But I got a fair bit done. I was in a pretty good mood when I got home.

I was boiling some spaghetti for dinner when my phone dinged.

Free?

Yes.

K. So tell me.

I thought everybody had low vit D.

Ha ha. No. Not really. I mean it's a bit complicated.

I feel like that should be the title of your autobiography. "It's a bit complicated."

How much detail do you want me to get into?

An excruciating amount of detail.

No detail is too small or obscure

Also, get really deep into the statistics.

Make sure you explore every possible nuance.

And don't stop even if I fall asleep.

Are you making fun of me?

It worries me you were uncertain enough to ask.

I tried to find the emoji of the smiling face sticking his tongue out, but my water boiled over and sizzled as it splashed on the stovetop.

I grabbed some oven mitts and moved the pot away before turning down the heat.

sorry, dinner ready and I only have two hands

It's okay. Log on when you sit down.

She texted me a long link, which I realized was for a video call. I tried to come up with a plausible excuse, but every possibility sounded weak and contrived. I didn't know why the prospect should make me feel anxious, but it did.

I strained and plated the spaghetti and looked down at my t-shirt. I went into the bathroom, looked at myself in the mirror and combed my hair. If it made a difference, I couldn't tell. I debated changing, but putting on a dress shirt seemed a bit much. Don't overthink it, I told myself. In the end, I put on a clean polo.

When I clicked on the link, Casey was sitting cross-legged on her couch in sweatpants and a hoodie with a lethargic Corgi resting in her lap.

"Hi! This is Rufus." She held up Rufus. He looked quizzically at his human and yawned. He put his head back on Casey's lap. She scratched behind his ears.

"He seems tired," I said.

"He's got a hectic schedule. Don't you Rufus?!?"

Rufus opened his eyes but decided against engaging and closed them again. Rufus had life figured out. "He doesn't seem interested in our conversation about vitamin D."

"He's a dog who eats trash he finds on the street if I don't watch him. He doesn't care much about nutrition or have the most discerning palate."

"That's fair."

"I on the other hand —"

"— *don't* eat trash you find on the street?"

Casey gave me a withering look. "No."

"That's reassuring."

"Anywaaaaaay, unlike Rufus here I do worry about vitamin D because my friend Julie just won't shut up about it. She says we all need to start taking vitamin D because you need it for bone strength and also to boost your immune system and fight off cancer and keep your arteries clean."

"People market it hard. I've heard all that stuff before."

"So we're not all vitamin D deficient?"

"Not by a long shot."

"But some people *do* have vitamin D deficiency, don't they?"

"Some, but not the majority of the population. Maybe 5 to 10 percent of people."

"I thought it was like half. And it's worse for us in North America because we don't spend much time outdoors, especially in the winter."

"You don't spend much time outdoors? Weather's a lot warmer where you are."

"If I go out in the sun, I burn like a tomato. Even if I put on sunscreen, I start getting freckles everywhere. Can you imagine me with freckles all over my body?"

I could, but I was trying to focus on the conversation.

"How come you're saying it only affects 5 to 10 percent of people and these websites say it's like half."

I shrugged. "Some of them are probably trying to sell vitamin D to people."

"So, what? They inflate the problem just so they can sell you the solution? . . . Never mind, I just answered my own question."

Part of the problem in defining vitamin D deficiency is the ongoing debate about what the normal level should be. Pick a higher threshold for normal, and more people get labeled deficient. If I were a cynical man, which I am, then I would think some people deliberately picked high thresholds to make the problem seem worse than it is.

"So what's normal?" Casey asked. "Just in case I want to go out and get tested."

"Most guidelines set it at 50 nanomoles per liter."

"Do I need to know what a nanomole is for this conversation to make sense?"

"Not really."

"Excellent. Proceed."

"Anyway, the point of medicine is to prevent disease —"

"I thought the point of medicine was to make lots of money, be passive-aggressive to patients, and write illegibly."

"That too. But sometimes prevent disease. If your vitamin D levels are above 50, you're probably not going to get sick. You won't get rickets."

"That's good. I don't want to end up like Tiny Tim."

"There's some debate about what disease Tiny Tim had. Some doctors argue it wasn't rickets. That he really had renal tubular acidosis."

"Do the doctors debating this issue realize he's a fictional character?"

"I hope so."

"Are you going to tell me what renal tubular acidosis is?"

"It's a kidney disease that can affect your bones and it can be corrected with a proper diet. That's why people think Tiny Tim got better after Scrooge started helping out the family and —"

Casey held up her hand. "Don't go professor mode on me. It's fine. I get it. Unless . . . do you have a tweed jacket with leather elbow patches? And can you go put it on right now?"

"I don't own anything like that."

Casey frowned. "That's very disappointing. Anyway, it doesn't matter. Point is, keep your vitamin D level above 50 thingamajiggers in your blood and you're fine. Right?"

I told Casey she was right, and she congratulated herself for her scientific prowess. If your serum vitamin D level is about 50 nanomoles then you're not deficient. There are certain people who think the threshold should be 75 nanomoles, but the general consensus is 50. Under 50 and you get labeled as having vitamin D insufficiency. Under 30 and you have vitamin D deficiency. Vitamin D deficiency is associated with worse health outcomes especially when it comes to bone health. The 30–50 range of insufficiency is more of a gray zone.

Casey nodded along but she periodically asked Rufus if he had any follow-up questions. I wasn't sure how seriously she was taking the conversation. Rufus had no questions. He snored. She pushed back a few times on why I kept underestimating the prevalence of vitamin D deficiency relative to the many social media influencers who had massive online followings.

Many people believe vitamin D deficiency is common and widespread. But the data doesn't back that up. It varies a bit from study to study and between countries. It can also change over time. But more or less somewhere around 5 to 10 percent of the population is truly deficient with roughly another 20 percent in the gray zone between 30 and 50 nanomoles per liter.

Casey pointed out that if you did the math, about a third of the population is on the lower side, which she didn't think was a trivial percentage. Granted that meant two-thirds of the population were fine. According to her, it came down to whether you were a glass-half-full or glass-half-empty type of person. But really, she just wanted to know if she needed to take a vitamin D supplement.

"I can't give you medical advice over the internet," I said. "I obviously don't know your medical history and all that."

"Plus you would obviously need to examine me first, right?"

I thought I saw a crack of a smile, but it was gone a nanosecond later. If she was toying with me, she was playing it straight. I wasn't sure what I was supposed to say so I changed the subject. "There is one group who should get routine vitamin D supplementation."

"Oh, who?"

"Do you want to guess?"

"Do I want to play a guessing game? Yes! I obviously want to play a guessing game. Sometimes I feel like you don't know me at all."

"Okay, so who —"

"What do I win if I guess right?"

"Umm . . ." I hadn't really thought this through. I looked around my apartment for inspiration but realized shipping her random objects from my home wasn't really going to be very practical.

"I'm waiting." She drummed her fingers on the keyboard.

My grandfather once told me that when you don't know what to do, you should always go with the obvious solution. The classics, after all, are classics for a reason.

"What's your favorite flower?" I asked.

Whatever Casey was expecting me to say, it clearly wasn't that. "Chrysanthemums."

"If you guess the correct answer, I'll send you an arrangement of chrysanthemums."

I'm not sure I can adequately describe the change that came over Casey at that point. Happiness? Yes. But also a steely-eyed gaze of fierce competitive intensity that was both intimidating and frightening. She really wanted those flowers.

"How many guesses do I get?" Rufus opened his eyes and raised his head to see what was going on. Why had the gentle rhythmic scratching behind his ears stopped?

"Three."

I had barely gotten out the words before Casey blurted out, "Old people!"

"I think you need to be more specific. You can't just guess the upper third of the human population."

"If there were rules and conditions on what I was allowed to guess, those needed to be explained prior to the commencement of guessing," Casey said. "We have an oral agreement and those are legally binding."

"I don't think that's true."

"Yeah, I don't know. I get most of my legal knowledge from *Law and Order*."

"It's not that. Older people might get prescribed vitamin D because they have osteoporosis or kidney failure or some other medical problem, but routinely giving it out to everyone over age 65 isn't a thing."

"Hmmm . . . okay. So what group needs to take vitamin D pills regularly?" Casey asked herself. "Ah. Vegans!"

"No, or at least not really. Vegetarians and vegans can become vitamin D deficient if they avoid specific foods but they aren't necessarily vitamin deficient."

"I thought because they avoid dairy and we get most of our vitamin D from milk."

"Except milk doesn't have vitamin D in it."

"I feel like all those public service announcements I saw as a kid lied to me."

Milk and dairy products have been the traditional vehicle of vitamin D supplementation, but contrary to popular belief, milk does not normally have vitamin D in it. It took a while to convince Casey. The vitamin D in milk is added. Starting in the 1930s and 1940s governments in the U.S. and Canada added it to combat diseases like rickets. Given that you no longer see children walking around with rickets, the program has been successful.

But you don't need to drink milk to get vitamin D. Fatty fish like salmon or mackerel, egg yolks, and liver are all good sources of vitamin D. Except, as Casey pointed out, they aren't really options for people who care about animals and just want to do right by Bambi.

It was a particular issue with milk substitutes. Many of them weren't supplemented with vitamin D and people who switched from cow milk to something like almond milk were at potential risk for vitamin deficiency, which struck Casey as unfair. Many people didn't realize government regulations for vitamin D supplementation in dairy products didn't apply to plant-based milks. Fortunately, things are a bit better now for people who want to avoid animal products. Orange juice, cereals, and plant-based milk products often have vitamin D added. It's not mandatory under the law, but it has become increasingly common, and a quick check of the product label will confirm it.

"I didn't realize grocery shopping required homework," Casey said.

I shrugged. "You're not automatically vitamin D deficient just because you're vegan. You just have to be a little bit careful to make sure you get vitamin D from alternative food sources."

"Can't you also make it from the sun?"

"You can. When sunlight hits your skin, your body makes vitamin D from cholesterol."

"Can't we just rely on that?"

"Not really. Your ability to make vitamin D decreases with age."

"One more reason not to get old then. So should I start standing by a window to soak up the sunshine and make my own vitamin D?"

"You can't get it through a window."

"Well that's just poop." Rufus raised himself up on his hind legs and licked her neck. "No. Down. Mommy's talking to people." After some cajoling, Rufus desisted and, distracted by something off-camera, he wandered off. "He saw a bird fly by the window," Casey said, "so he's going to go do a security sweep of the perimeter."

"He must be a good guard dog."

"It's his house. He just lets me live here with him. What do you mean you can't get it through a window?"

"It's not sunlight that triggers the production of vitamin D. It's the sun's UV radiation."

"So? UV radiation goes through glass."

"Actually, no it doesn't."

"That just blew my mind. Seriously?"

"Haven't you ever thought about why the people in radiology departments are always standing behind glass when you go for an X-ray or a CT scan or whatever?"

"Never really thought about it to be honest with you. I wouldn't have thought a greenhouse would be the safest place to be in a nuclear blast."

Though I'm not a nuclear physicist, I was pretty sure it wasn't. One has to assume the concussive force of the explosion would almost certainly break the glass and shower tiny shards down on you. Casey saw the problem and decided we should just avoid nuclear war altogether. I wholeheartedly agreed. Casey looped back to her earlier point. "So glass blocks UV radiation? Well, that's a mind-bender. So I can't get sunburned or a tan through glass either?"

"Nope. If you have a sunroom or a solarium in your home —"

"You vastly overestimate how nice my place is."

"Then let's say you're lying on your couch and the afternoon sun is coming through your window."

"That sounds like my typical Sunday afternoon."

"You'll be warm and toasty, but your skin won't burn. Which is good, but you won't make any vitamin D either."

"If you need UV rays to make it, doesn't that mean sunscreen would block your body's ability to make vitamin D?"

It would. Casey thought it was an unfair trade-off. But skin cancer is pretty bad and easy to prevent, and you can get all the vitamin D you need from food. Since fish exist, you can just eat fish. I didn't think the skin damage from the sun was worth it in the long term. Not just because of the cancer risk but also because of photoaging. Most of what we perceive visually as "getting older" is photodamage to the skin brought on by sun exposure. People who avoid the sun and wear sunscreen have fewer wrinkles and look younger.

Casey sat up and took notice. "You're saying a $20 bottle of sunscreen is going to prevent wrinkles?"

"Yes."

"Geez, you're just chock-full of useful tidbits. I'm going to keep you around. I need to learn stuff like this. Plus I need someone who can write me prescriptions for painkillers and sleeping pills and stuff." The alarm must have showed on my face because Casey laughed. "You are so easy to mess with."

I didn't take the bait. I told her she still had a third guess.

She did an admirable job of stalling. She asked Rufus for advice, but he was of no help. She tried to ply some hints out of me, but when I wouldn't budge, she finally guessed "kids." It was a very broad guess, and also wrong. True, many places have school milk programs, but that's more about breakfast rather than specifically a vitamin D supplementation scheme. She finally gave up.

"Newborns," I told her. "Specifically, newborns who are exclusively breastfed. It's recommended that they get a vitamin D supplement."

"Just breastfed babies? Why?"

"Because breast milk doesn't have very much vitamin D in it."

"That feels like a major design flaw."

"It does beg a series of existential questions. But it's true. Vitamin D does not get expressed in breast milk."

"Why are bottle-fed babies left out of this?" Casey took a moment and answered her own question. "Oh. Because they put vitamin D directly into the formula, right?"

"Exactly."

Casey thought for a moment. "So I was right when I said young people."

"Your guess felt a little vague to me. I don't think that qualifies for the grand prize."

"Well, I think we need to review that decision because —" She craned her neck to look at something off-camera. "I think I have to take Rufus for a walk. He's getting a little fidgety."

"Definitely not something you want to put off."

"Nope. I don't need to clean up that mess." She said bye and the screen went to black.

I looked down at my plate. I'd barely eaten any of my food.

I didn't hear from Casey again for a few days. I was tempted to text her, but I worried she might be busy, and I couldn't think of anything clever to say either. Inspiration came to me one afternoon. I found a florist online and ordered some chrysanthemums. They gave me the option to add a card, so I typed in my message. "Close Enough." I didn't include my name. I was pretty confident I wouldn't have to.

I heard from Casey two days later. I figured it would take at least that long for the flowers to be delivered, so I wasn't too surprised.

She texted a picture of the flowers on her windowsill bathed in the morning sunlight. Rufus lay on the ground paws up in the middle of a rectangle of sunlight.

I'm glad you decided to honor our agreement. 😍 Thank you!

I'm not sure Rufus likes the flowers though.

For him, that response is a very strong endorsement. Plus he feels a lot better about lying in the sun ever since you told him the window is blocking UV rays.

BTW, how come plants still grow when you put them by a window?

Cause photosynthesis in plants depends on light, not UV rays.

Rufus, Herbert, and I thank you.

I was about to ask who Herbert was, but I got a call to go see a consult in the emergency room. I collected my stethoscope, put some extra pens in my pocket and headed downstairs. As I walked toward the ER, I tried to remember if Casey had ever mentioned anyone named Herbert. I was pretty sure she said she didn't have a roommate. Maybe it was her brother. Or a cousin. Or a boyfriend. Maybe sending flowers was a mistake.

I stewed on it as I slogged my way through patients. However, I didn't get to where I am without the ability to compartmentalize. At around my sixth consultation for high blood pressure, I decided to break for lunch. I thought about Herbert as I ate my sandwich. When in doubt, best to play it cool. And brevity is always the soul of wit. A one-word query was the way to go.

Herbert?

Herbert is the chrysanthemum. I named him.

I hope you give him a good home

He seems really happy here.

Plus he and Rufus seem to really get along.

Largely because Herbert is high enough on the windowsill that Rufus can't realistically pee in the flowerpot.

Which would have really damaged their relationship I think.

I laughed and typed "LOL." I returned to the emergency room in a decidedly better mood.

The remainder of my week on call was uneventful. The days were rather long, and I usually got home tired. I ate, showered, slept, and I strategically ignored my emails. One afternoon, Casey sent a string of text that I missed.

So, it's not that I don't trust you

But I looked some stuff up online after we talked.

And a lot of people say that taking vit D is good for bones

But also helps fight cancer and prevents heart disease.

And boosts the immune system and lots of other stuff

So you and the internet can't both be right.

You're busy doing doctor stuff aren't you?

Oh well. Go save a life or something.

I'll make up a list of questions.

When I got home that night, I decided to give Casey proof of life.

Sorry it was a crazy week.

There was no reply. I mindlessly scrolled through social media. But I didn't have to wait long.

There's some really weird medical stuff online.

People feel very strongly about vitamin D

Like weirdly passionate about it.

Kind of scary

Also, I forgive you for ignoring me.

Were you saving lives?

I hope so.

Okay. Well. Party's over. Back to vitamin D.

Why do people think it cures cancer?

People do. Doctors don't.

Some doctors do. I saw them online.

Are they also trying to sell you vitamin D?

Looks like it.

But they gotta make a living

Lake-front vacation homes aren't cheap.

I imagine.

Maybe I'll find out one day. **sigh**

A person's reach must exceed their grasp

Otherwise, what's heaven for?

I like that you randomly quote Robert Browning.

But you never answered my question.

Why do people think that vitamin D cures cancer?

Or reduces the risk of getting cancer. Whatever

It has to do with a concept called confounding.

Is this a researcher/statistical thing?

Sort of.

Will there be graphs?

I might have to resort to some basic diagrams, yes.

SWITCHING TO VIDEO!!!

Before I had finished reading the all-caps response, the video call came in.

"Hi!" Casey squinted. "Are you in bed already?"

"Just about."

"Are those flannel pajamas? That is the most adorable thing I have ever seen. Do you also have the childhood teddy bear you grew up with sitting on a bureau by your bed because you can't bring yourself to get rid of him?"

I looked at the bear. "No, of course not. I'm a grown man."

"Hmmm . . . sure. Okay. So you said something about confounding?"

This was going to be a bit complicated. Early research showed a link between low vitamin D levels and chronic diseases like cancer or heart disease. But those early findings weren't borne out by randomized trials. There was an explanation, but I thought it might be easier to explain with an analogy.

"Let's say aliens came to Earth to do medical studies on humans."

"That might be the only way I get comprehensive medical coverage."

I declined to comment. "Imagine these aliens wanted to find out why humans get lung cancer. What would they do?"

"Based on TV, they would rely heavily on rectal probing."

"In practice, that would probably not help."

"What if they're alien gastroenterologists and they're trying to start a galaxy-wide colon cancer screening program?"

I had my doubts that Casey was taking this example seriously. But I was committed to seeing the analogy through to the end. "The aliens are here to study lung cancer. The logical thing for them to do is to go around and question people to see what makes lung cancer patients different from everyone else."

"How did they learn English?"

"I don't know . . . they learned English beforehand."

"A bunch of intergalactic alien polyglot epidemiologists came to Earth to study lung cancer. That's our starting premise?"

"It is. Now, they discover lung cancer patients are more likely to carry lighters in their pockets. So they assume, not knowing anything about human culture —"

"They don't know anything about human culture, but they know how to speak English?"

"You're just messing with me now, aren't you?"

Casey giggled.

"So not knowing anything about human culture, they deduce that carrying a lighter in your pocket leads to lung cancer," I said.

"Except, a lighter in and of itself doesn't cause lung cancer."

This was the critical point. "Exactly, because there's a third element that influences the relationship between the other two."

"You lost me."

I knew this would be complicated. "Okay, time for a diagram."

Casey cheered.

"Hold on, let me get a piece of paper." I was sort of comfy lying on my bed, but I got up and grabbed a notebook from my desk. Back in bed, I sketched a graph and held it up to the camera.

"That sir is a triangle."

"For this analogy to make sense —"

"We need wine. On it." She disappeared from the screen and came back a few minutes later, drink in hand. "Proceed."

"We have to begin by agreeing that lighters do not cause cancer in and of themselves."

"Whatever floats your boat, Captain." Casey took a long sip. "But that's obviously true. That's not how lighters work … presumably. I don't actually know how they work when you get right down to it."

To be honest, I didn't know how lighters worked either. But that didn't matter. The point was smoking linked those two variables together. Lighters don't cause cancer. But smokers carry lighters, and smokers are more likely to get lung cancer. Casey pointed out a few potential problems with my example. Smokers don't absolutely have to carry lighters. They can carry matches or bum a light off someone else. And not all smokers get lung cancer. They were fair points, but it's true that smokers were more likely to carry lighters and get lung cancer than the general population. It only looks like lighters are linked to lung cancer because there's a backdoor path going from lighters to smokers to cancer, hence my dotted arrow.

I was proud of my analogy, but Casey said it was pretty simple, almost insultingly plebeian for a woman of her breeding and sophistication. Hints of the British accent were coming out.

I took it up a level and talked about hormone replacement therapy. In the 1990s, hormone replacement therapy was routinely given to postmenopausal women. It was once thought that it could prevent chronic diseases like heart disease or cancer but its use dropped off after the publication of the Women's Health Initiative study. Before its publication, the research suggested a cardiac and cancer benefit. After, everybody realized the early data was wrong. As Casey put it, it came crashing down like when you play a drunken game of Jenga with your friends. I had no frame of reference to judge the accuracy of the statement.

I made another sketch to show her why that happened.

"Same as before," I said. "Hormone replacement therapy, HRT, looks like it's linked to heart disease because there's a third variable, lifestyle, joining them together."

"What do you mean lifestyle?"

"It's a grab bag term for diet, exercise, socioeconomic status."

"Socioeconomic status?"

"A measure of wealth. When you have disposable income you can buy healthier food, you have more free time to exercise, and you can buy better medical care. A lot of things work in your favor."

"Damn it. I knew being poor was a bad idea."

"See what you can do about that."

"I'm doing my best!! Believe it or not, that Nigerian prince didn't want to share his fortune with me. And he was really uncool about the whole thing when I started asking follow-up questions."

"You answered the email?"

"Sometimes I get bored at night and start trolling online scammers. It's how I bring a little bit of sanity back into the world."

I couldn't keep a straight face. Casey had a habit of being unpredictable and that kept me on my toes. I liked that. It made things interesting. And as I was thinking all this, Casey reminded me what we were supposed to be talking about.

The downfall of hormone replacement therapy was caused by a simple fact. The women who went out and got hormone prescriptions from their doctors were healthier to begin with and more health-conscious than those who didn't. Health-conscious women sought out hormone therapy. And because they were health-conscious at baseline, they were at lower risk for heart disease. Their baseline health status and lifestyle factors made hormone therapy seem cardio-protective. But it wasn't.

Casey said she got it but also wanted to know what any of this had to do with vitamin D. She said she wasn't in a hurry, as she had plenty of wine, but she was worried it might be past my bedtime and that my teddy bear might be missing me.

I ignored the jab. "Same thing happened with vitamin D."

"So it's like this?" Casey drew something on a piece of paper and held it up to the camera.

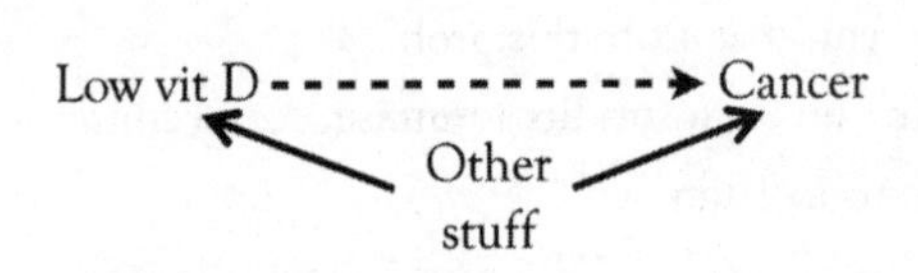

"Got it in one. Older people are more likely to be vitamin D deficient and more likely to get cancer. Another one is a poor diet —"

"Because if you don't eat properly, you're more likely to be vitamin D deficient and more likely to get cancer."

"That's it. Vitamin D deficiency looks bad if you see it. But giving somebody vitamin D doesn't fix the underlying problem."

"The underlying problem being the relentless march of time and the inherent frailty of the human condition?"

I stared back blankly.

She tried again. "Vitamin D can't fix the fact that you got old or eat mostly junk?"

I told her that it can't. Vitamin D for all its promise was no time machine. Casey was a bit surprised researchers couldn't account for these baseline differences statistically. To some degree you can, but you can never account for the differences perfectly. You can adjust for someone's age in your statistical analysis, but someone's birthday doesn't perfectly reflect their state of health. I have patients in their 90s who take hour-long walks every day, still labor in their gardens, and are in complete possession of their faculties, and I have patients in their 60s who can't say the same.

Other factors are even harder to account for. Participants in studies don't always tell you if they smoke. Diet is a hard thing to record in a spreadsheet. There are a lot of qualitative characteristics that define the human condition. Not everything can be dealt with statistically.

"So teenage me was right," Casey said.

I told her I didn't understand.

"Everything is stupid and nothing really matters."

I took a page from Casey's book. "Womp, womp."

"Don't steal my bit!" She took another sip of wine.

"The only real solution to this problem, the only way to completely eliminate confounding, is to do a randomized trial."

I have great admiration for people who conduct randomized trials. They are big, complicated, expensive logistical nightmares. The statistical analysis is usually pretty easy. The planning is hard. But it has that one great advantage. If you take a group of people and randomly assign them to vitamin D or placebo, then everything else should balance out. Because the group assignments are completely random, you should have the same number of diabetics, smokers, or cancer survivors in each group. If the numbers are large enough, the differences equalize.

History has shown us why we need randomized trials. If all we have is observational data, then hormone replacement therapy looks like it's good for your heart. Once you do the randomized trial, you realize it isn't. Same thing happened with vitamin D. We had mountains of observational studies linking vitamin D deficiency to everything from high blood pressure to diabetes to heart disease, to autoimmune problems, to cancer.

There was a time when vitamin D seemed like the cure for everything. Casey thought somebody should have pointed out that this was too good to be true. She was probably right. But a lot of people got swept up in the enthusiasm, including me.

"What changed your mind?" Casey asked.

"The randomized trials."

First we had a large meta-analysis of 21 randomized trials where vitamin D didn't lower the risk of heart disease. Then we had the VITAL study, which looked at vitamin D's effects on both heart disease and cancer, specifically breast, prostate, and colon cancer. Neither improved with vitamin D. In the AMATERASU study, vitamin D didn't extend survival in patients with gastrointestinal cancer. In the D2d study, it didn't prevent diabetes, and in the COPSAC study, it didn't prevent asthma in children when given to pregnant mothers. Finally, in the VITAL-DKD study, it didn't prevent kidney damage if you had diabetes.

Casey listened, without interrupting, until I got to the end of the list. "So I'm kind of impressed you were able to rattle those off from memory."

"I had to review it all for a summary article, so it kind of stuck. It was a year-in-review piece."

"All those studies came out in the same year?!?"

"2019 was a bad year for vitamin D enthusiasts."

Casey considered it for a minute and absent-mindedly petted Rufus, who had returned to the place of honor in her lap. "Let me play devil's advocate. Maybe the studies failed because the people in them weren't deficient."

"They've tried that. The VDKA study tried vitamin D to treat asthma in kids who were deficient. They gave them a massive dose of 4,000 units per day. Nothing."

"Well that's just poop."

"They tried it specifically in the elderly because they're more likely to be deficient. That was the DO-HEALTH study, and it came back negative."

Casey held up her hand. "I get it, you're just beating a dead horse here."

"It drives me a bit nuts that they've studied vitamin D for so many possible diseases in so many possible scenarios and it keeps coming back negative. At a certain point, you start crossing the line from optimism into delusion."

"So it does nothing. I'm very disappointed. I like it when stories have happy endings. This is depressing."

"It still has value in treating osteoporosis and ensuring proper skeletal health."

"That's something. But nothing beyond the bone stuff?" Rufus leaped to attention and barked. "Rufus! No! Down! Mommy's on the phone. There's nothing to eat. Go sleep in your bed." Rufus looked around, saw that there was no bone on offer, and rolled over on his back for a belly rub.

"Sorry about that," Casey said. "Certain words sometimes set him off. Whatever you do, don't say C-H-I-C-K-E-N."

"Duly noted. But to answer your question. Even the bo—" I caught myself. "Even the skeletal benefits of vitamin D are a bit uncertain. There was recently a study where they gave high doses of vitamin D to healthy adults. They gave them either 4,000 or 10,000 units *per day*, and the usual dose is 10,000 units per week. They did that for years and it didn't improve bone density. Unless you actually have osteoporosis or some specific medical problem, it doesn't look as if vitamin D does anything."

"But maybe by the time you're an adult, it doesn't matter anymore. Maybe it makes the biggest difference if you give it to kids cause they're still growing. Like Tiny Tim."

"Intuitively, that would make sense."

"You're about to tell me that's been studied before, aren't you?"

"I am. It was in Mongolia where there's a lot of vitamin D deficiency. They took children and gave them high doses of vitamin D every week for three years to see if it improved growth development."

"If it's going to work for anyone, it's going to be in that group," Casey said.

I said nothing.

"It didn't work, did it?"

"It's kind of amazing that it didn't, but it didn't."

"That's very counterintuitive."

"That's why you have to test things in medicine. If you don't test it, you won't know. Vitamin D seems like it should work, but it doesn't unless you have specific medical problems. Hormone replacement therapy looks like it should work, but it doesn't. That's confounding. Until you do a proper randomized trial, you can easily get fooled."

"The first principle is that you must not fool yourself, and you're the easiest person to fool."

Not many people know that quote. "Richard Feynman. Nice."

"I'm an impressive girl, what can I say?" Casey gestured theatrically. "So I guess I'll tell Julie that it's not so clear-cut for vitamin D. Maybe for bone health but —" Rufus barked. "Oh for goodness' sake. Hold on. I'm going to go give him some food."

She was only gone for a minute, but my eyelids got heavy and I came close to nodding off.

"I'll let you get to sleep," she said.

"It's close to 2 a.m." I yawned. "I better turn in."

"Night night. Pleasant dreams."

"You too. Sweet dreams."

I was in good spirits the next morning, until I got to my office. One of our colleagues was out sick and that set off a chain reaction that threw the entire medical schedule into disarray. It may be physically impossible to be in two places at once, but the on-call schedule said different. I neglected emails and messages for the better part of the week, relying on Irene to flag anything urgent.

It was the end of a long day, at the end of a long week, when I peeked up from an increasingly cluttered desk. Alexi stood in my doorway. I'd forgotten he was due back, and it was already 8 p.m. He wanted to go for dinner. We went to a small hamburger place around the corner.

"You never answered me when I texted you," he said.

"If it's any consolation, I ignored a lot of other texts." I told him Jim had messaged. He was coming to town, and he and Katie were going on some sort of spa retreat.

He raised an eyebrow. "He sure knows how to punch above his weight class."

"Fortune favors the bold."

"Speaking of which, whatever happened to that barista? I really liked her. She had your number. You should go visit her."

"That's a bit presumptuous?"

"I'm always amazed you can't follow your own advice."

"And what advice is that?"

"You need to be bold. Or at least 10 percent less timid. That girl liked you. You should have asked her out."

"There are practical difficulties to living in different cities."

"Jim figured it out. You know why?"

"Enlighten me."

"I'm sure he's a nice guy, but he strikes me as the type of person who's too oblivious to realize he might fail. He has just enough self-confidence to paper over his many other deficiencies. They invented the term *failing upward* with him in mind."

"Harsh."

"Do you know why he has a weekend planned with Katie?"

"Why?"

"Because he asked her. That's why. You miss 100 percent of the shots you don't take."

"I like it when you speak in hockey metaphors."

"You should have gotten the barista's number."

It told him I'd actually spoken to her.

He paused mid-bite. "Okay! Good! See, the system works. I lowered my expectations and now I'm pleasantly surprised things worked out the way they were supposed to."

He wanted details. I didn't know what that meant, so I said nothing.

He put down his burger. "God, you're annoying. What did you talk about?"

"Vitamin D."

He seemed confused. "I don't . . . is that some pop culture reference I don't recognize?"

"No, we were talking about vitamin D. The vitamin."

"In what context?"

"In the medical context." I was confused by his confusion.

Alexi was silent. "I don't believe you. No one could possibly be that . . ." He grabbed my phone and unlocked it. I'd been using the same password for 15 years. He tapped and swiped with increasing intensity. "Jesus! You weren't kidding." He looked at me. "How long have you been texting? I don't speak to my girlfriend this much!"

"Fiancée."

"Don't change the subject!" He kept scrolling. "There's a time stamp here from a video call that says you spoke for nearly three hours."

In my defense, confounding is a complicated concept to explain. Alexi looked like he might scream, and he threatened to knock some sense into me. Instead, he said he was going to tell me what to do. Ask her out. Fly to LA for a weekend. Sit down and have a meal with her. "Talk about normal things that people care about."

"The vitamin and supplement industry is a billion-dollar business. People care about it. Trust me."

Alexi put the phone down and ran his hands through his hair. "Okay. Listen to me very carefully. Because this is important."

I told him I was all ears, but I was making a mental note to pick up milk and some tomatoes on the way home.

He tapped the phone. "Casey doesn't care about confounding. Or selection bias. Or about medical statistics in general. She isn't genuinely worried about whether she needs to take a vitamin D supplement."

"She specifically asked about vitamin D because a friend mentioned to her —"

"She cares about it in a general sense. But she doesn't care about the minutiae of medical epistemology and methodological research practices. I know you think people find this stuff interesting. But they don't." He held up a hand. "Don't get me wrong, it's good people like you exist to keep the rest of us honest. And I'm not saying this stuff isn't vitally important for preserving the integrity of medical research. But it's not the reason Casey is texting you at 1 a.m. It's not quite 'u up?' but she's not that curious about the state of affairs in medical academia."

"I don't know what 'u up?' means."

Alexi rubbed his temples. "Are you putting me on right now? I actually can't tell if you're messing with me."

"We just enjoy talking with each other. That's all this is."

Alexi looked baffled and stood up.

"Where are you going?"

"I'm going to the bathroom. I'm enraged. I need to pee."

"That might be correlation rather than causation."

"Don't start!"

I thought about what he said. But I didn't think he was right. Casey had her own life. Plus it didn't really make sense to try anything long-distance. Those things only work in movies.

I looked down at my phone and saw messages I had yet to read, and a string of messages from Casey that I had read but never answered, because I kept drafting and discarding replies.

Are you planning to come back to California for any conferences soon?

I have lots more questions for you if you can come by the coffee shop again. LOL

Rufus also wants to meet you.

He has several questions about bones LOL

I'm sure he means osteoporosis, but maybe stop by the butcher's before you come over.

She'd sent several photos of Rufus. He was, I had to admit, photogenic. I was impressed that Casey managed to get him to sit still long enough to put glasses on his nose, pose a book in front of him, and get the picture. He did seem especially hungry for knowledge. Or soup bones.

It would be nice to see Casey again. Maybe we could go for a walk and bring Rufus and sit on a park bench and drink hot chocolate now that the weather was getting colder. I could probably make a long weekend trip as well if I planned my schedule right.

I was being silly. I had no reason to go to LA. There were no major conferences coming up. The cost of an airline ticket and hotel would be, not exactly prohibitive, but still considerable. And it would be a lot of money to . . . what? . . . ask someone on a date?

I typed a message.

Sadly, no planned conferences coming up and very hard to get time off work these days. But we can talk whenever you like. Happy to answer any science questions you have. That's what I'm here for.

I reviewed the message. It was the only sensible answer. I didn't have any plans to travel again. One conference a year is more than enough, especially now that material is posted online anyway. There was a minor conference on communicating science to the public coming up, but I didn't know anybody who would be there. Plus, I don't like to fly.

Alexi came back to the table. "Are you ready?"

"No, not really, but might as well do it anyway."

"What was that?"

"Sorry, nothing." I told him I needed to finish a reply then we could go grab a drink to celebrate his return.

I deleted what I'd written and tried again.

I'll be there next month. Do you want to have dinner while I'm in town? . . . It will be good to see you. 😀

I hit send without rereading the message.

Epilogue

"Everything's got a moral, if only you can find it."

— LEWIS CARROLL

I told you this wasn't a diet book.

If you want to eat healthy, it's simple. Potato chips are bad for you. No amount of statistical contortionism can make that not true. Same goes for french fries, ice cream, and chocolate chip cookies. Life is unfair like that. Fruits, vegetables, and whole grains are better, but I probably don't need to tell you that. You know all this already. It's just that sometimes someone needs to remind us of the fundamentals. Because if you watch the news or doom-scroll the internet, you'll think it's much more complicated than this. It isn't. It's simple. The internet makes it complicated because a lot of the stories you read about food are just not true.

Hot dogs aren't as bad for you as cigarettes. You might agree with Jim that bacon tastes awesome, but you probably shouldn't eat meat every day. You probably don't need to avoid eggs completely like Katie, but you should still care about cholesterol. And it's really much more about the cholesterol in your blood than the cholesterol in your food.

You shouldn't see wine as health food or vitamin D as the cure for everything. You should eat less salt because it increases your blood pressure. But there's no number to aim for. At least none we can agree on. You should eat less salty food because that stuff isn't good for you to begin with. It probably doesn't really matter if you skip breakfast. But if you do have a morning cup of coffee, you can rest assured it's not going to give you cancer or cause a heart attack. But if it makes you all fluttery like Casey, then maybe you should cut back. Caffeine is a stimulant.

The trick is understanding why people get this stuff wrong. The reasons are often subtle, mathematical, and nuanced. And you can't fit much nuance into 140 characters. But put yourself in the shoes of researchers, and you quickly realize why food questionnaires are blunt instruments. Analyze your data to death and you'll probably find *something* you can publish, but there's a difference between perseverance and random statistical noise. Small risks can look big if you frame them the right way. Numbers don't lie, but they can be misleading.

You shouldn't lie to other people, and you definitely shouldn't lie to yourself. If you want to go celebrate something with your friends, I'm not going to tell you you can't. They're your friends. I don't know them. You can have wine if you want, just don't drive afterward. And do it because you want to, not because you think it's good for your heart. But maybe drink less. Alcohol is a carcinogen after all.

The problem is old myths die slow deaths. We're still hauling outdated scientific baggage into our public consciousness. Vitamin C doesn't do much against the common cold, and vitamin D won't cure cancer. Dark chocolate isn't healthy. We just don't understand satire anymore.

The core principles still hold true, and you know what those are. You know how to eat right. You don't need me to tell you. That's why this isn't a diet book. This is a love story.

Acknowledgments

Writing a book is a very solitary process. But publishing a book takes a whole lot of people to make it happen.

First off, I want to thank Robert Lecker, my agent. He took my idea for a book and turned it into a plan. Given that I knew nothing about the literary world before starting the process, his guidance was instrumental in getting the project started.

There would be no book without everyone at ECW Press. In particular, Jack David was willing to take a chance on a first-time author. Our early conversations really helped shape what this book ultimately became. He suggested it should be a series of conversations between characters. Once he said it, I saw the potential.

While this is my first book (second if you count a technical manual on echocardiography for cardiology trainees), I've spent eight years doing a podcast with my very good friend Jonathan Jarry called *The Body of Evidence*. Although our on-screen personas bicker, fight, and drive each other crazy, we have one of the best working relationships I've ever known. We've spent eight years producing a bonkers, off-the-wall medical comedy podcast involving time travel, a teddy bear, and a car-driving dog. It shouldn't work, but it does. So when it came time to write a book that weaved humor and science together, I had plenty of practice. He is a good friend and one of the best professional partners I've ever had. He is, however, still a poopy head.

The "real" Casey read an early draft of the first chapter involving her namesake. She sent back a single-line email which said, "Pretty good. But I kept expecting him to ask her out at the end." Until she said it, I didn't realize I was writing a romantic comedy.

Every good writer needs a good editor. I have no idea if I am a good writer, but I had a good editor. Lesley Erickson took on the difficult task of editing the first draft of a first-time author. If the book is remotely readable, it is thanks to her. Jen Albert also went through the manuscript and caught many repetitions, continuity errors, and general weak spots. I hear that some writers don't like it when editors pick at their work. I have no such compunction. I am indebted to both of them because they made the book better. Samantha Chin helped this book along and guided this novice through a process he barely understood. Her patience has been infinitely appreciated.

Many of my friends helped, directly and indirectly, to populate this book with interesting characters. In particular, my friend Wanda read every chapter and provided early editorial feedback. Her early support and encouragement were vital. Marlena unintentionally inspired a character with a single offhand comment, "Yuck." A random dinner with Alexi led to a whole new character and two major chapters. Jason and Robin endured many questions from me and helped me work out many of the little details that made the characters real. I love both Bevs tremendously, but a long-standing running joke about bacon with one of them wove its way through this book. Lest anyone think these characters are based on real people, they are not. They are works of fiction and the products of my imagination. But my friends inspired some of them because my friends inspired me. They helped me through the pandemic and through the writing process. Writing a book is a solitary process. But no man who has friends is alone.

References and Further Reading

PREFACE

The graph where every food both causes and prevents cancer was adapted from a study in the *American Journal of Clinical Nutrition*.

Schoenfeld, Jonathan D., and John P.A. Ioannidis. "Is Everything We Eat Associated with Cancer? A Systematic Cookbook Review." *American Journal of Clinical Nutrition* 97, no. 1 (January 1, 2013): 127–34. https://doi.org/10.3945/ajcn.112.047142.

CHAPTER 1: VITAMIN C FIGHTS THE COMMON COLD

All the studies looking at Vitamin C in terms of the prevention or treatment of the common cold are discussed and reviewed in the 2013 Cochrane review discussed in this chapter. Anyone willing to read through the 90-page document will find a detailed and comprehensive discussion about all the different studies done on this subject over the past several decades:

Hemilä, Harri, and Elizabeth Chalker, "Vitamin C for Preventing and Treating the Common Cold," *Cochrane Database of Systematic*

Reviews, January 31, 2013. https://doi.org/10.1002/14651858.CD000980.pub4.

The ISIS-2 study which looked at how the benefits of aspirin varied by astrological sign can be found here:

ISIS-2 (Second International Study of Infarct Survival) Collaborative Group. "Randomized Trial of Intravenous Streptokinase, Oral Aspirin, Both, or Neither among 17,187 Cases of Suspected Acute Myocardial Infarction: ISIS-2." *Lancet* 2, no. 8607 (August 13, 1988): 349–60.

The Miracle DICE study is also worth reading. It is funny and clever and original and makes the same important point that if you go looking for patterns you will find them, even if they don't exist:

Counsell, Carl E., Mike J. Clarke, Jim Slattery, and Peter A.G. Sandercock. "The Miracle of DICE Therapy for Acute Stroke: Fact or Fictional Product of Subgroup Analysis?" *BMJ* 309, no. 6970 (December 24, 1994): 1677–81. https://doi.org/10.1136/bmj.309.6970.1677.

Many people have covered the Brian Wansink story, including some of the earliest reporting by Stephanie Lee of BuzzFeed:

Lee, Stephanie M. "Here's How Cornell Scientist Brian Wansink Turned Shoddy Data into Viral Studies about How We Eat." Accessed March 1, 2023. https://www.buzzfeednews.com/article/stephaniemlee/brian-wansink-cornell-p-hacking.

Although Wansink's original blog post is gone, an archived version of it is still available online:

"The Grad Student Who Never Said 'No'—Healthier & Happier."

Accessed March 2, 2023. https://web.archive.org/web/20170312041524/http:/www.brianwansink.com/phd-advice/the-grad-student-who-never-said-no.

For anyone who wants a more in-depth explanation of what p-values and statistical significance actually mean, *Vox* has a good explainer on the subject:

Resnick, Brian. "800 Scientists Say It's Time to Abandon 'Statistical Significance.'" *Vox*. March 22, 2019. https://www.vox.com/latest-news/2019/3/22/18275913/statistical-significance-p-values-explained.

The humorous graphs showing spurious correlations were taken from Tyler Vigen's blog "Spurious Correlations." Accessed March 2, 2023. https://www.tylervigen.com/spurious-correlations.

The regulation of vitamins, supplements, and natural health products is a complex issue that could probably justify a book in itself. For anyone who wants a quick summary of this issue, John Oliver did a fairly thorough and simultaneously humorous take on the subject here:

Oliver, John. "Dr. Oz and Nutritional Supplements." *Last Week Tonight with John Oliver* (HBO). June 23, 2014. https://www.youtube.com/watch?v=WAowKeokWUU.

CHAPTER 2: HOT DOGS ARE AS BAD AS CIGARETTES

Anyone wanting to learn more about IARC and how they do what they do can consult their official website:

IARC. "About IARC." Accessed March 2, 2023. https://www.iarc.who.int/cards_page/about-iarc/.

The classification scheme they use can also be found on their website and explains how and what criteria are used to put agents into Group 1, 2a, 2b or 3:

IARC. "Classified by the IARC Monographs Volumes 1–132 – IARC Monographs on the Identification of Carcinogenic Hazards to Humans." Accessed March 4, 2023. https://monographs.iarc.who.int/agents-classified-by-the-iarc/.

The story of caprolactam, the only agent to temporarily be listed in Group 4, can be found in various news stories of the day:

Loria, Kevin. "Yoga Pants Contain the Only Known Substance That 'Probably Does Not Cause Cancer.'" *Business Insider*, May 27, 2016. https://www.businessinsider.com/almost-everything-causes-cancer-2016-5.

Anyone willing to do so can read the entire 517-page IARC review on red meat and processed red meat by downloading it from the IARC website here:

IARC. *Red Meat and Processed Meat: IARC Monographs on the Evaluation of Carcinogenic Risks to Humans Volume 114*. 2016. Accessed March 2, 2023. https://publications.iarc.fr/Book-And-Report-Series/Iarc-Monographs-On-The-Identification-Of-Carcinogenic-Hazards-To-Humans/Red-Meat-And-Processed-Meat-2018.

However, anyone not interested in perusing the over 800 studies described in the text can review some of the major studies described in this chapter:

The Women's Health Initiative

Beresford, Shirley A.A., Karen C. Johnson, Cheryl Ritenbaugh, Norman L. Lasser, Linda G. Snetselaar, Henry R. Black, Garnet L. Anderson, et al. "Low-Fat Dietary Pattern and Risk of Colorectal Cancer: The Women's Health Initiative Randomized Controlled Dietary Modification Trial." *JAMA* 295, no. 6 (February 8, 2006): 643–54. https://doi.org/10.1001/jama.295.6.643.

Cancer Prevention Study II Nutrition Cohort

Chao, Ann, Michael J. Thun, Cari J. Connell, Marjorie L. McCullough, Eric J. Jacobs, W. Dana Flanders, Carmen Rodriguez, Rashmi Sinha, and Eugenia E. Calle. "Meat Consumption and Risk of Colorectal Cancer." *JAMA* 293, no. 2 (January 12, 2005): 172–82. https://doi.org/10.1001/jama.293.2.172.

The EPIC Cohort

Norat, Teresa, Sheila Bingham, Pietro Ferrari, Nadia Slimani, Mazda Jenab, Mathieu Mazuir, Kim Overvad, et al. "Meat, Fish, and Colorectal Cancer Risk: The European Prospective Investigation into Cancer and Nutrition." *Journal of the National Cancer Institute* 97, no. 12 (June 15, 2005): 906–16. https://doi.org/10.1093/jnci/dji164.

The Nurses Health Study Cohort

Willett, W.C., M.J. Stampfer, G.A. Colditz, B.A. Rosner, and F.E. Speizer. "Relation of Meat, Fat, and Fiber Intake to the Risk of Colon Cancer in a Prospective Study among Women." *The New England Journal of Medicine* 323, no. 24 (December 13, 1990): 1664–72. https://doi.org/10.1056/NEJM199012133232404.

The recommendation to continue eating red meat came from the NutriRECS Consortium:

Johnston, Bradley C., Dena Zeraatkar, Mi Ah Han, Robin W.M. Vernooij, Claudia Valli, Regina El Dib, Catherine Marshall, et al. "Unprocessed Red Meat and Processed Meat Consumption: Dietary Guideline Recommendations from the Nutritional Recommendations (NutriRECS) Consortium." *Annals of Internal Medicine* 171, no. 10 (November 19, 2019): 756–64. https://doi.org/10.7326/M19-1621.

The conflict-of-interest issue surrounding the NutriRECS consortium was detailed by journalists including Laura Reiley at the *Washington Post*:

Reiley, Laura. "Research Group That Discounted Risks of Red Meat Has Ties to Program Partly Backed by Beef Industry." *Washington Post*, October 23, 2019. https://www.washingtonpost.com/business/2019/10/14/research-group-that-discounted-risks-red-meat-has-ties-program-partly-backed-by-beef-industry/.

The lifetime risk of getting colorectal cancer is calculated as 4.2 percent for males and 4.0 percent for females based on the 2022 Cancer Statistics from the American Cancer Society. For simplicity, I just listed the statistic as 4 percent for everyone. The IARC report estimated that eating an extra 50 grams of processed meat per day increased your risk of colorectal cancer by 18 percent; in other words, it was 1.18 times higher. So a lifetime risk of colorectal cancer of 4 percent that is 1.18 times higher works out to a 4.72 percent lifetime risk of colorectal cancer, which I rounded up to 5 percent, again for simplicity. So eating meat every day increases your risk of getting colorectal cancer by a little bit under 1 percent if you are at average risk to begin with.

Siegel, Rebecca L., Kimberly D. Miller, Hannah E. Fuchs, and Ahmedin Jemal. "Cancer Statistics, 2022." *CA: A Cancer Journal for*

Clinicians 72, no. 1 (January 2022): 7–33. https://doi.org/10.3322/caac.21708.

The "Random Medical News" cartoon by Jim Borgman, first published by the *Cincinnati Inquirer* and King Features Syndicate, April 27, 1997; Forum section: 1, and reprinted in the *New York Times*, April 27, 1997, E4.

We use a large amount of antibiotics in animals although the amount does appear to be decreasing because of increasing awareness of antibiotic resistance. We actually use more antibiotics in animals than humans overall, but the differences are less when you account for the greater biomass (i.e. the total weight) of animals used in farming. The European Medicines Agency actually uses a different metric called the Population Correction Unit (PCU) to monitor antibiotic use in animals. All this to say, antibiotic use in animals and humans is roughly comparable and is also changing as new legislation limits the non-medical use of antibiotics in animals.

Moulin, Gérard, Philippe Cavalié, Isabelle Pellanne, Anne Chevance, Arlette Laval, Yves Millemann, Pierre Colin, Claire Chauvin, and on behalf of the 'Antimicrobial Resistance' ad hoc Group of the French Food Safety Agency. "A Comparison of Antimicrobial Usage in Human and Veterinary Medicine in France from 1999 to 2005." *Journal of Antimicrobial Chemotherapy* 62, no. 3 (September 1, 2008): 617–25. https://doi.org/10.1093/jac/dkn213.

GOV.UK. "Understanding the Mg/PCU Calculation Used for Antibiotic Monitoring in Food Producing Animals." Accessed March 2, 2023. https://www.gov.uk/government/publications/understanding-the-mgpcu-calculation-used-for-antibiotic-monitoring-in-food-producing-animals.

Van Boeckel, Thomas P., Emma E. Glennon, Dora Chen, Marius Gilbert, Timothy P. Robinson, Bryan T. Grenfell, Simon A. Levin,

Sebastian Bonhoeffer, and Ramanan Laxminarayan. "Reducing Antimicrobial Use in Food Animals." *Science* 357, no. 6358 (September 29, 2017): 1350–52. https://doi.org/10.1126/science.aao1495.

CHAPTER 3: SOME SALT IS GOOD FOR YOU

You can find many articles written about salt, both supporting and criticizing the argument that we should eat less of it. The *Scientific American* article that opens the chapter is as good a place as any to understand why some people are deeply skeptical and critical of recommendations to eat less salt:

Moyer, Melinda Wenner. "It's Time to End the War on Salt." *Scientific American*, July 8, 2011. https://www.scientificamerican.com/article/its-time-to-end-the-war-on-salt/.

The issues of people underreporting their weight and overreporting their height are real and predictable to the point that correction factors for the values people report on questionnaires are available for researchers to use:

Shields, Margot, Sarah Connor Gorber, Ian Janssen, and Mark S. Tremblay. "Bias in Self-Reported Estimates of Obesity in Canadian Health Surveys: An Update on Correction Equations for Adults." *Health Reports* 22, no. 82 (2011).

The debate of how to best measure salt intake is a complex one that often requires trading off accuracy for simplicity. In the end, all the different strategies have advantages and disadvantages:

Cook, Nancy R., Feng J. He, Graham A. MacGregor, and Niels Graudal. "Sodium and Health — Concordance and

Controversy." *BMJ (Clinical Research Ed.)* 369 (June 26, 2020): m2440. https://doi.org/10.1136/bmj.m2440.

Food questionnaires are one of the more common tools used in food research despite the many issues discussed in the chapter. The actual food questionnaire developed by Harvard is copyrighted, but anyone who wants to try filling out an example of one can do so at:

"Nutrition Questionnaire Service Center." Harvard T.H. Chan School of Public Health, Department of Nutrition. Accessed March 2, 2023. https://www.hsph.harvard.edu/nutrition-questionnaire-service-center/.

Salt remains particularly challenging because so much of the salt we eat comes from food we do not prepare ourselves:

He, Feng J., and Graham A. MacGregor. "Reducing Population Salt Intake Worldwide: From Evidence to Implementation." *Progress in Cardiovascular Diseases* 52, no. 5 (2010): 363–82. https://doi.org/10.1016/j.pcad.2009.12.006.

Measuring salt in urine samples is not without its own issues. Accuracy varies depending on what type of urine collection you use. In a re-analysis of the TOHP study, researchers showed that when you compared 24-hour urine specimens and spot urine samples, you got different results. Using 24-hour urine collections produced the usual results that lower sodium levels reduced deaths. But when they used spot urine specimens, they found low sodium levels correlated with higher death rates. The authors concluded that the observation that low sodium levels increased cardiovascular risk was a statistical anomaly caused by information bias and inaccurate sodium measurements from spot urine collections:

He, Feng J., Norm R.C. Campbell, Yuan Ma, Graham A. MacGregor, Mary E. Cogswell, and Nancy R. Cook. "Errors in Estimating

Usual Sodium Intake by the Kawasaki Formula Alter Its Relationship with Mortality: Implications for Public Health." *International Journal of Epidemiology* 47, no. 6 (December 1, 2018): 1784–95. https://doi.org/10.1093/ije/dyy114.

Some of the major studies discussed in this chapter are listed below but they represent only a small snapshot of the research on salt. Anybody interested in reading about the body of evidence on salt and its effect on human health should consider reading *Salt Wars* by Michael Jacobson.

Jacobson, Michael F. *Salt Wars: The Battle over the Biggest Killer in the American Diet*. Cambridge, Massachusetts: The MIT Press, 2020.

Some of the key studies looking at the association between salt intake and health outcomes include the following:

The INTERSALT study looked at the association between salt intake and blood pressure:

"Intersalt: An International Study of Electrolyte Excretion and Blood Pressure. Results for 24 Hour Urinary Sodium and Potassium Excretion. Intersalt Cooperative Research Group." *BMJ (Clinical Research Ed.)* 297, no. 6644 (July 30, 1988): 319–28. https://doi.org/10.1136/bmj.297.6644.319.

The DASH diet tested whether less sodium could reduce blood pressure:

Sacks, F.M., L.P. Svetkey, W.M. Vollmer, L.J. Appel, G.A. Bray, D. Harsha, E. Obarzanek, et al. "Effects on Blood Pressure of Reduced Dietary Sodium and the Dietary Approaches to Stop Hypertension (DASH) Diet. DASH-Sodium Collaborative Research Group." *The New England Journal of Medicine* 344, no. 1

(January 4, 2001): 3–10. https://doi.org/10.1056/NEJM20010104344 0101.

The Trials of Hypertension Prevention (TOHP) tested the cardiovascular benefits of salt reduction:

Cook, Nancy R., Jeffrey A. Cutler, Eva Obarzanek, Julie E. Buring, Kathryn M. Rexrode, Shiriki K. Kumanyika, Lawrence J. Appel, and Paul K. Whelton. "Long Term Effects of Dietary Sodium Reduction on Cardiovascular Disease Outcomes: Observational Follow-up of the Trials of Hypertension Prevention (TOHP)." *BMJ (Clinical Research Ed.)* 334, no. 7599 (April 28, 2007): 885–88. https://doi.org/10.1136/bmj.39147.604896.55.

These findings were upended by the PURE study, which found an increase in cardiovascular risk with a very low sodium diet:

Mente, Andrew, Martin J. O'Donnell, Sumathy Rangarajan, Matthew J. McQueen, Paul Poirier, Andreas Wielgosz, Howard Morrison, et al. "Association of Urinary Sodium and Potassium Excretion with Blood Pressure." *The New England Journal of Medicine* 371, no. 7 (August 14, 2014): 601–11. https://doi.org/10.1056/NEJMoa1311989.

The bitterness and increasing polarization of the "Salt Wars" as they came to be called was summed up in an editorial in the *International Journal of Epidemiology*:

Ioannidis, John P.A. "Commentary: Salt and the Assault of Opinion on Evidence." *International Journal of Epidemiology* 45, no. 1 (February 2016): 264–65. https://doi.org/10.1093/ije/dyw015.

The Institute of Medicine acknowledged some of the shortcomings of the existing data in their 2013 report:

McGuire, Shelley. “Institute of Medicine. 2013. Sodium Intake in Populations: Assessment of Evidence. Washington, DC: The National Academies Press, 2013.” *Advances in Nutrition* 5, no. 1 (January 4, 2014): 19–20. https://doi.org/10.3945/an.113.005033.

But the proposed solution of a study on prisoners ended up being even more controversial. The editorial announcing the plan and the rationale behind it was published in the journal *Hypertension*:

Jones, Daniel W., Friedrich C. Luft, Paul K. Whelton, Michael H. Alderman, John E. Hall, Eric D. Peterson, Robert M. Califf, and David A. McCarron. “Can We End the Salt Wars With a Randomized Clinical Trial in a Controlled Environment?” *Hypertension* 72, no. 1 (July 2018): 10–11. https://doi.org/10.1161/HYPERTENSIONAHA.118.11103.

That editorial now contains a correction. Originally no conflicts of interest were declared. Only later did the ties between one of the researchers and the food industry come to light. One of the many articles detailing this story was done by Stephanie Lee, formerly of *BuzzFeed News*:

Lee, Stephanie. “A Prison Study Aims To End The ‘Salt Wars.’ It Turns Out The Salt Industry Wants To Help Fund It.” *BuzzFeed News*, September 18, 2018. https://www.buzzfeednews.com/article/stephaniemlee/salt-institute-sodium-study-prison-funding.

The name of the group as the “Jackson 6” was used by David McCarron in an interview he gave after the publication of the editorial:

Husten, Larry. “Salt War Opponents Unite in Call for Randomized Trial in Prisons.” *CardioBrief*, May 21, 2018. http://www.cardiobrief.org/2018/05/21/salt-war-opponents-unite-in-call-for-randomized-trial-in-prisons/.

While many people might be shocked that the United States was experimenting on prisoners during World War II, they probably shouldn't be. The experiments were not a secret:

"Prison Malaria: Convicts Expose Themselves to Disease So Doctors Can Study It," *Life*, June 4, 1945, 43–46.

Today, research on prisoners is much more tightly regulated. While it is possible to do research in prisons, it has to be justified in much clearer terms. Anyone interested in learning the criteria one has to meet to begin this type of research project can consult the Health and Human Services website:

HHS.gov. "Prisoner Research FAQs." Accessed March 2, 2023. https://www.hhs.gov/ohrp/regulations-and-policy/guidance/faq/prisoner-research/index.html.

The SSaSS study which tested a 75/25 sodium/potassium salt did show a reduction in cardiac events, which seems to support the idea that we should be eating less salt:

Neal, Bruce, Yangfeng Wu, Xiangxian Feng, Ruijuan Zhang, Yuhong Zhang, Jingpu Shi, Jianxin Zhang, et al. "Effect of Salt Substitution on Cardiovascular Events and Death." *The New England Journal of Medicine* 385, no. 12 (September 16, 2021): 1067–77. https://doi.org/10.1056/NEJMoa2105675.

But the SODIUM-HF trial which tried to reduce sodium intake in heart failure patients showed no benefit:

Ezekowitz, Justin A., Eloisa Colin-Ramirez, Heather Ross, Jorge Escobedo, Peter Macdonald, Richard Troughton, Clara Saldarriaga, et al. "Reduction of Dietary Sodium to Less than 100 Mmol in Heart Failure (SODIUM-HF): An International, Open-Label,

Randomized, Controlled Trial." *Lancet* 399, no. 10333 (April 9, 2022): 1391–1400. https://doi.org/10.1016/S0140-6736(22)00369-5.

CHAPTER 4: COFFEE CAUSES CANCER

The *Time* magazine tweets about coffee and the sometimes funny comments that followed them are available online. Nothing ever dies on the internet.

TIME. "How Coffee Can Help You Live Longer." Twitter, December 26, 2016. https://twitter.com/TIME/status/813214025551122432.

TIME. "The Problem with Your Coffee." Twitter, January 8, 2017. https://twitter.com/TIME/status/818064770964324352.

The 1981 *NEJM* paper about coffee and pancreatic cancer has become a classic case study on selection bias:

MacMahon, B., S. Yen, D. Trichopoulos, K. Warren, and G. Nardi. "Coffee and Cancer of the Pancreas." *The New England Journal of Medicine* 304, no. 11 (March 12, 1981): 630–33. https://doi.org/10.1056/NEJM198103123041102.

MacMahon's interview with the *EpiMonitor* illustrates his thinking at the time of publication:

Epidemiology Monitor. "Coffee and Pancreatic Cancer: An Interview with Brian MacMahon." April–May 2001. Accessed March 3, 2023. https://www.epimonitor.net/Reprint_of_the_Month.htm.

Although ultimately other researchers found issues with some of those conclusions and published their own data which showed no link whatsoever:

Feinstein, Alvan R., Ralph I. Horwitz, Walter O. Spitzer, and Renaldo N. Battista. "Coffee and Pancreatic Cancer: The Problems of Etiologic Science and Epidemiologic Case-Control Research." *JAMA* 246, no. 9 (August 28, 1981): 957–61. https://doi.org/10.1001/jama.1981.03320090019020.

Goldstein, H.R. "No Association Found between Coffee and Cancer of the Pancreas." *The New England Journal of Medicine* 306, no. 16 (April 22, 1982): 997. https://doi.org/10.1056/NEJM198204223061625.

Wynder, E.L., N.E. Hall, and M. Polansky. "Epidemiology of Coffee and Pancreatic Cancer." *Cancer Research* 43, no. 8 (August 1983): 3900–3906.

Gold, E.B., L. Gordis, M.D. Diener, R. Seltser, J.K. Boitnott, T.E. Bynum, and D.F. Hutcheon. "Diet and Other Risk Factors for Cancer of the Pancreas." *Cancer* 55, no. 2 (January 15, 1985): 460–67. https://doi.org/10.1002/1097-0142(19850115)55:2<460::aid-cncr2820550229>3.0.co;2-v.

Hsieh, C.C., B. MacMahon, S. Yen, D. Trichopoulos, K. Warren, and G. Nardi. "Coffee and Pancreatic Cancer (Chapter 2)." *The New England Journal of Medicine* 315, no. 9 (August 28, 1986): 587–89.

The *New York Times* article that introduced the public to this potential cancer link between coffee and cancer was written by Harold M. Schmeck Jr. and published on March 12, 1981. Three months later, the second article refuted that claim.

Harold M. Schmeck, Jr. "Study Links Coffee Use to Pancreas Cancer." *The New York Times*, March 12, 1981. https://www.nytimes.com/1981/03/12/us/study-links-coffee-use-to-pancreas-cancer.html.

Harold M. Schmeck, Jr. "Critics Say Coffee Study Was Flawed." *The New York Times*, June 30, 1981. https://www.nytimes.com/1981/06/30/science/critics-say-coffee-study-was-flawed.html.

The *LA Times* wrote a similar account in 1986 after the second part of MacMahon's study was published:

L.A. Times Archives. "Developments in Brief: Drink Up: New Research Disputes Reported Link Between Coffee, Cancer." *Los Angeles Times*, August 31, 1986. https://www.latimes.com/archives/la-xpm-1986-08-31-me-15067-story.html.

The story of Barry Marshall and the discovery of *H. pylori* is a fascinating one and has been documented in some of the many articles and interviews that he has given over the years:

Marshall, Barry, and Paul C. Adams. "Helicobacter Pylori: A Nobel Pursuit?" *Canadian Journal of Gastroenterology* 22, no. 11 (November 2008): 895–96.

Weintraub, Pamela. "The Doctor Who Drank Infectious Broth, Gave Himself an Ulcer, and Solved a Medical Mystery." *Discover Magazine*, April 8, 2010. https://www.discovermagazine.com/health/the-doctor-who-drank-infectious-broth-gave-himself-an-ulcer-and-solved-a-medical-mystery.

The examples of selection bias involving NBA players, undergrads, and dating are all variations on the same theme and are often used in teaching to explain what can be a complex topic. Anybody interested in learning more can read about these examples in some of the key epidemiological texts on the subject.

Glymour, M. Maria. "Using Causal Diagrams to Understand Common Problems in Social Epidemiology." In *Methods in Social*

Epidemiology, edited by J. Michael Oakes and Jay S. Kaufman, 393–428. San Francisco: Jossey-Bass, 2006.

The California court case involving coffee was a long, drawn-out saga that made international headlines.

Associated Press. "Judge Rules Coffee Sold in California Needs Cancer Warnings." *CBC News*, March 29, 2018. https://www.cbc.ca/news/health/california-coffee-cancer-warnings-1.4599906.

The lawsuit was brought using Proposition 65, the California statute requiring businesses to provide warning labels on products that contain carcinogenic ingredients.

California Environmental Protection Agency Office of Environmental Health Hazard Assessment. "About Proposition 65." OEHHA, January 12, 2015. https://oehha.ca.gov/proposition-65/about-proposition-65.

But ultimately California's Office of Environmental Health and Hazard Assessment (OEHHA) declared that no warning label was required and in 2020 the case was dismissed.

California Environmental Protection Agency Office of Environmental Health Hazard Assessment. "Final Statement of Reasons Title 27, California Code of Regulations Adoption of New Section 25704 Exposures to Listed Chemicals in Coffee Posing No Significant Risk." Accessed March 3, 2023. https://oehha.ca.gov/media/downloads/crnr/fsorcoffee060719.pdf.

Randazzo, Sara. "Coffee Doesn't Warrant a Cancer Warning in California, Agency Says." *Wall Street Journal*, June 3, 2019, sec. Business. https://www.wsj.com/articles/coffee-doesnt-warrant-a-cancer-warning-in-california-agency-says-11559604158.

The IARC monograph linking coffee to cancer was actually more concerned with hot beverages than coffee specifically.

IARC Working Group on the Evaluation of Carcinogenic Risks to Humans. *Drinking Coffee, Mate, and Very Hot Beverages*. IARC Monographs on the Evaluation of Carcinogenic Risks to Humans. Lyon (FR): International Agency for Research on Cancer, 2018. http://www.ncbi.nlm.nih.gov/books/NBK543953/.

CHAPTER 5: RED WINE'S GOOD FOR YOUR HEART

The origins of the French Paradox stem from some early observations trying to explain why heart disease rates in France were lower than in other countries.

Renaud, S., and M. de Lorgeril. "Wine, Alcohol, Platelets, and the French Paradox for Coronary Heart Disease." *Lancet* 339, no. 8808 (June 20, 1992): 1523–26. https://doi.org/10.1016/0140-6736(92)91277-f.

Simini, B. "Serge Renaud: From French Paradox to Cretan Miracle." *Lancet* 355, no. 9197 (January 1, 2000): 48. https://doi.org/10.1016/S0140-6736(05)71990-5.

But the idea really only permeated into the public consciousness after it was featured on *60 Minutes* with Morley Safer. You can still see the original segment online:

"How Morley Safer Convinced Americans to Drink More Wine," *CBS News*, August 28, 2016. https://www.cbsnews.com/news/how-morley-safer-convinced-americans-to-drink-more-wine/.

However, not everyone accepted the idea that red wine was responsible for the low rates of heart disease in France and alternative explanations were put forth at the time.

Artaud-Wild, S.M., S.L. Connor, G. Sexton, and W.E. Connor. "Differences in Coronary Mortality Can Be Explained by Differences in Cholesterol and Saturated Fat Intakes in 40 Countries but Not in France and Finland. A Paradox." *Circulation* 88, no. 6 (December 1993): 2771–79. https://doi.org/10.1161/01.cir.88.6.2771.

Law, Malcolm, and Nicholas Wald. "Why Heart Disease Mortality Is Low in France: The Time Lag Explanation." *BMJ* 318, no. 7196 (May 29, 1999): 1471–80.

It's worth remembering that the 1980s were a different time when cardiac risk factors were treated far less aggressively than they are now. In the 1986 Oslo Hypertension Study, patients were treated for "mild" hypertension that ranged between 150 and 180 mmHg. These numbers would be considered fairly high by today's standards.

Leren, P., and A. Helgeland. "Oslo Hypertension Study." *Drugs* 31 Suppl 1 (1986): 41–45. https://doi.org/10.2165/00003495-198600311-00008.

While resveratrol has some biological activity in the lab, studies like the InCHIANTI study have cast doubt about whether it has any effect in humans. It also seems unlikely that humans get enough of it from food to make a difference, even from grapes and red wine. You would probably have to drink gallons of red wine every day to get enough resveratrol into your system, a certainly lethal amount.

Weiskirchen, Sabine, and Ralf Weiskirchen. "Resveratrol: How Much Wine Do You Have to Drink to Stay Healthy?" *Advances in*

Nutrition 7, no. 4 (July 11, 2016): 706–18. https://doi.org/10.3945/an.115.011627.

Semba, Richard D., Luigi Ferrucci, Benedetta Bartali, Mireia Urpí-Sarda, Raul Zamora-Ros, Kai Sun, Antonio Cherubini, Stefania Bandinelli, and Cristina Andres-Lacueva. "Resveratrol Levels and All-Cause Mortality in Older Community-Dwelling Adults." *JAMA Internal Medicine* 174, no. 7 (July 2014): 1077–84. https://doi.org/10.1001/jamainternmed.2014.1582.

Another argument against resveratrol is that studies have shown the same benefit with not just red wine but also other types of alcohol, where resveratrol concentrations are markedly less.

Klatsky, A.L., M.A. Armstrong, and G.D. Friedman. "Red Wine, White Wine, Liquor, Beer, and Risk for Coronary Artery Disease Hospitalization." *The American Journal of Cardiology* 80, no. 4 (August 15, 1997): 416–20. https://doi.org/10.1016/s0002-9149(97)00388-3.

Rimm, E.B., A. Klatsky, D. Grobbee, and M.J. Stampfer. "Review of Moderate Alcohol Consumption and Reduced Risk of Coronary Heart Disease: Is the Effect Due to Beer, Wine, or Spirits?" *BMJ (Clinical Research Ed.)* 312, no. 7033 (March 23, 1996): 731–36. https://doi.org/10.1136/bmj.312.7033.731.

People who drink wine are on average wealthier and in better health than those who don't according to studies like this one from Denmark. This would suggest that the health benefits seen in wine drinkers have more to do with the fact that they are healthier to begin with, not so much because wine itself is having any positive impact.

Mortensen, E.L., H.H. Jensen, S.A. Sanders, and J.M. Reinisch. "Better Psychological Functioning and Higher Social Status May

Largely Explain the Apparent Health Benefits of Wine: A Study of Wine and Beer Drinking in Young Danish Adults." *Archives of Internal Medicine* 161, no. 15 (August 13, 2001): 1844–48. https://doi.org/10.1001/archinte.161.15.1844.

Individuals are very bad at measuring the appropriate amount of alcohol. In this study, college students were asked to free-pour drinks and they consistently poured too much.

White, Aaron M., Courtney L. Kraus, Julie D. Flom, Lori A. Kestenbaum, Jamie R. Mitchell, Kunal Shah, and H. Scott Swartzwelder. "College Students Lack Knowledge of Standard Drink Volumes: Implications for Definitions of Risky Drinking Based on Survey Data." *Alcoholism, Clinical and Experimental Research* 29, no. 4 (April 2005): 631–38. https://doi.org/10.1097/01.alc.0000158836.77407.e6.

Part of the confusion probably comes from the fact that guidelines in the U.S., Canada, and the UK are different and what constitutes a standard drink changes based on what country you happen to be in.

Rethink Your Drinking. "What's a Standard Drink?" Accessed March 3, 2023. https://www.rethinkyourdrinking.ca/what-is-a-standard-drink/.

National Institute on Alcohol Abuse and Alcoholism (NIAAA). "What Is A Standard Drink?" Accessed March 3, 2023. https://www.niaaa.nih.gov/alcohols-effects-health/overview-alcohol-consumption/what-standard-drink.

Drinkaware. "What Is an Alcohol Unit?" Accessed March 3, 2023. https://www.drinkaware.co.uk/facts/alcoholic-drinks-and-units/what-is-an-alcohol-unit.

Some researchers think current guidelines are too high and government recommendations about alcohol should be half as much as they currently are. This meta-analysis in the *Lancet* found that health risks associated with alcohol started to go up when people drank more than 100 grams of alcohol per week, which is roughly half the current recommendations for "moderate" consumption.

Wood, Angela M., Stephen Kaptoge, Adam S. Butterworth, Peter Willeit, Samantha Warnakula, Thomas Bolton, Ellie Paige, et al. "Risk Thresholds for Alcohol Consumption: Combined Analysis of Individual-Participant Data for 599 912 Current Drinkers in 83 Prospective Studies." *Lancet* 391, no. 10129 (April 14, 2018): 1513–23. https://doi.org/10.1016/S0140-6736(18)30134-X.

Recent Canadian guidelines recommended a much lower threshold. After they reviewed the data, they found that health risks start going up after two drinks per week. The risk is proportional and higher the more you drink. But their fundamental point is that no level of alcohol consumption is really good for you.

Paradis, C., P. Butt, K. Shield, N. Poole, S. Wells, T. Naimi, A. Sherk, and the Low-Risk Alcohol Drinking Guidelines Scientific Expert Panels. "Canada's Guidance on Alcohol and Health: Final Report." Canadian Centre on Substance Use and Addiction, January 2023. https://ccsa.ca/sites/default/files/2023-01/CCSA_Canadas_Guidance_on_Alcohol_and_Health_Final_Report_en.pdf.

Unfortunately, people who value the potential health benefits of red wine choose to ignore the many negative health consequences of alcohol. Namely the risk for many types of common cancers including breast cancer. Also, alcohol can increase the risk of postoperative complications.

Cao, Yin, Walter C. Willett, Eric B. Rimm, Meir J. Stampfer, and Edward L. Giovannucci. "Light to Moderate Intake of Alcohol, Drinking Patterns, and Risk of Cancer: Results from Two Prospective US Cohort Studies." *BMJ* 351 (August 18, 2015): h4238. https://doi.org/10.1136/bmj.h4238.

Eliasen, Marie, Marie Grønkjær, Lise Skrubbeltrang Skov-Ettrup, Stine Schou Mikkelsen, Ulrik Becker, Janne Schurmann Tolstrup, and Trine Flensborg-Madsen. "Preoperative Alcohol Consumption and Postoperative Complications: A Systematic Review and Meta-Analysis." *Annals of Surgery* 258, no. 6 (December 2013): 930–42. https://doi.org/10.1097/SLA.0b013e3182988d59.

Cutting back on alcohol can lead to a number of health benefits. Not only does it help lower blood pressure but it can help decrease certain arrhythmias like atrial fibrillation.

Roerecke. M., J. Kaczorowski, S.W. Tobe, G. Gmel, O.S.M. Hasan, and J. Rehm. "The Effect of a Reduction in Alcohol Consumption on Blood Pressure: A Systematic Review and Meta-analysis." *Lancet Public Health* (February 2017): e108–e120. https://doi.org/10.1016/S2468-2667(17)30003-8. Epub. February 7, 2017.

Voskoboinik, A., J.M. Kalman, A. De Silva, T. Nicholls, B. Costello, S. Nanayakkara, S. Prabhu, D. Stub, S. Azzopardi, D. Vizi, et al. "Alcohol Abstinence in Drinkers with Atrial Fibrillation." *The New England Journal of Medicine* 382, no. 1 (January 2, 2020): 20–28.

Ronald Fisher probably does deserve his title as one of the fathers of modern statistics. But even geniuses can get things wrong and his insistence that cigarettes were not responsible for lung cancer was a remarkable blind spot. Anyone interested in delving deeper into this fascinating story can read Ben Christopher's essay on the topic.

Christopher, Ben. "Why the Father of Modern Statistics Didn't Believe Smoking Caused Cancer." *Priceonomics*, September 21, 2016. https://priceonomics.com/why-the-father-of-modern-statistics-didnt-believe/.

Multiple examples of reverse causation exist in the medical literature. Identifying them and teasing out the true statistical effects is not always easy.

Sattar, Naveed, and David Preiss. "Reverse Causality in Cardiovascular Epidemiological Research: More Common Than Imagined?" *Circulation* 135, no. 24 (June 13, 2017): 2369–72. https://doi.org/10.1161/CIRCULATIONAHA.117.028307.

The figure showing the difference in heart failure risk between former and never drinkers comes from an analysis by the Cardiovascular Health Study. The researchers still thought there might be some mild protective benefit at low levels of alcohol consumption. But the more interesting finding, in my opinion, is how different the risk was between former drinkers and never drinkers. Mixing them together could easily skew the association and make abstinence look much worse than it actually is.

Bryson, Chris L., Kenneth J. Mukamal, Murray A. Mittleman, Linda P. Fried, Calvin H. Hirsch, Dalane W. Kitzman, and David S. Siscovick. "The Association of Alcohol Consumption and Incident Heart Failure: The Cardiovascular Health Study." *Journal of the American College of Cardiology* 48, no. 2 (July 18, 2006): 305–11. https://doi.org/10.1016/j.jacc.2006.02.066.

One of the examples used in the chapter is the issue of smoking and mental illness. Unfortunately, people with mental illness have higher smoking rates than the general population.

National Institute on Drug Abuse. "Do people with mental illness and substance use disorders use tobacco more often?" National Institute on Drug Abuse website. Accessed March 3, 2023. https://nida.nih.gov/publications/research-reports/tobacco-nicotine-e-cigarettes/do-people-mental-illness-substance-use-disorders-use-tobacco-more-often.

Genetic studies like Mendelian Randomization studies suggest that alcohol doesn't have a cardioprotective benefit.

Millwood, Iona Y., Robin G. Walters, Xue W. Mei, Yu Guo, Ling Yang, Zheng Bian, Derrick A. Bennett, et al. "Conventional and Genetic Evidence on Alcohol and Vascular Disease Aetiology: A Prospective Study of 500 000 Men and Women in China." *Lancet* 393, no. 10183 (May 4, 2019): 1831–42. https://doi.org/10.1016/S0140-6736(18)31772-0.

Unfortunately, a formal randomized trial like the MACH study will likely not happen any time soon given that the NIH has withdrawn funding after the Advisory Committee to the Director published their report and made their recommendations.

"NIH to End Funding for Moderate Alcohol and Cardiovascular Health Trial," National Institutes of Health, June 15, 2018. https://www.nih.gov/news-events/news-releases/nih-end-funding-moderate-alcohol-cardiovascular-health-trial.

NIH Advisory Committee to the Director (ACD). "ACD Working Group for Review of the Moderate Alcohol and Cardiovascular Health Trial," NIH.gov, June 2018. https://acd.od.nih.gov/documents/presentations/06152018Tabak-B.pdf.

CHAPTER 6: CHOCOLATE IS HEALTH FOOD

The chocolate and Nobel Prize study was a real study published in the *New England Journal of Medicine.*

Messerli, Franz H. "Chocolate Consumption, Cognitive Function, and Nobel Laureates." *New England Journal of Medicine* 367, no. 16 (October 18, 2012): 1562–64. https://doi.org/10.1056/NEJMon1211064.

It generated many headlines and news stories like this one, which happily told people that chocolate could boost mental performance and increase your chances of winning a Nobel Prize.

"Eating Chocolate Produces Nobel Prize Winners, Says Study." Confectionery News. October 10, 2012. https://www.confectionerynews.com/Article/2012/10/11/Chocolate-creates-Nobel-prize-winners-says-study.

The "fake" chocolate story claiming that chocolate can help you lose weight was the brainchild of John Bohannon, a journalist who had previously exposed problems with open-access journals. The story of how he and his team designed the fake chocolate study can be read in an article he wrote for Gizmodo.

Bohannon, John. "I Fooled Millions into Thinking Chocolate Helps Weight Loss. Here's How." Gizmodo. May 27, 2015. https://gizmodo.com/i-fooled-millions-into-thinking-chocolate-helps-weight-1707251800.

The Reuters article about the study linking chocolate with Nobel Prizes contained many quotes that were obviously meant to be jokes but ended up being used by many other news outlets. In fact, the article itself clearly states that both Messerli and Cornell were joking and that they thought the correlation documented in the paper was absurd. A few articles did take a more critical eye though and realized both the intended satire of the piece and the issues it raised about drawing faulty conclusions.

Joelving, Frederik. "Eat Chocolate, Win the Nobel Prize?" *Reuters Health*, sec. Healthcare & Pharma. October 10, 2012. https://www.reuters.com/article/us-eat-chocolate-win-the-nobel-prize-idUSBRE8991MS20121010.

A similar article in *Popular Science* also took a critical look at the study:

Dillow, Clay. "The More Chocolate a Nation Eats, The More Nobel Prizes It Gets," *Popular Science*, October 12, 2012. https://www.popsci.com/science/article/2012-10/nations-chocolate-intake-directly-correlated-number-nobel-laureates-it-spawns-not-really/.

The statement issued by Cornell to the BBC radio show *More or Less*:

Pritchard, Charlotte. "Does Chocolate Make You Clever?" BBC News, November 16, 2012. https://www.bbc.com/news/magazine-20356613.

Some academics also critiqued the association:

Maurage, Pierre, Alexandre Heeren, and Mauro Pesenti. "Does Chocolate Consumption Really Boost Nobel Award Chances? The Peril of Over-Interpreting Correlations in Health Studies." *The Journal of Nutrition* 143, no. 6 (June 1, 2013): 931–33. https://doi.org/10.3945/jn.113.174813.

The Danish Diet, Cancer, and Health Study suggested a link between eating chocolate and a lower risk of atrial fibrillation. Data from the Women's Health Study supported that benefit, but data from the Physicians' Health Study did not.

Mostofsky, Elizabeth, Martin Berg Johansen, Anne Tjønneland, Harpreet S. Chahal, Murray A. Mittleman, and Kim Overvad. "Chocolate Intake and Risk of Clinically Apparent Atrial Fibrillation: The Danish Diet, Cancer, and Health Study." *Heart* 103, no. 15 (August 2017): 1163–67. https://doi.org/10.1136/heartjnl-2016-310357.

Conen, David, Stephanie E. Chiuve, Brendan M. Everett, Shumin M. Zhang, Julie E. Buring, and Christine M. Albert. "Caffeine Consumption and Incident Atrial Fibrillation in Women." *The American Journal of Clinical Nutrition* 92, no. 3 (September 2010): 509–14. https://doi.org/10.3945/ajcn.2010.29627.

Khawaja, Owais, Andrew B. Petrone, Yousuf Kanjwal, John Michael Gaziano, and Luc Djoussé. "Chocolate Consumption and Risk of Atrial Fibrillation (From the Physicians' Health Study)." *The American Journal of Cardiology* 116, no. 4 (August 15, 2015): 563–66. https://doi.org/10.1016/j.amjcard.2015.05.009.

The Cochrane review on the effect of cocoa on blood pressure found a small difference in blood pressure. The clinical significance of that difference is questionable.

Ried, Karin, Peter Fakler, and Nigel P Stocks. "Effect of Cocoa on Blood Pressure." *The Cochrane Database of Systematic Reviews* 2017, no. 4 (April 25, 2017): CD008893. https://doi.org/10.1002/14651858.CD008893.pub3.

The *Vox* article about how chocolate became a health food details how the industry started funding research on this topic.

Belluz, Julia. "Dark Chocolate Is Now a Health Food. Here's How That Happened." *Vox*, October 18, 2017. https://www.vox.com/science

-and-health/2017/10/18/15995478/chocolate-health-benefits-heart-disease.

The COSMOS study and COSMOS-Mind sub-study are the largest randomized trials on chocolate to date.

Sesso, Howard D., JoAnn E. Manson, Aaron K. Aragaki, Pamela M. Rist, Lisa G. Johnson, Georgina Friedenberg, Trisha Copeland, et al. "Effect of Cocoa Flavanol Supplementation for the Prevention of Cardiovascular Disease Events: The Cocoa Supplement and Multivitamin Outcomes Study (COSMOS) Randomized Clinical Trial." *The American Journal of Clinical Nutrition* 115, no. 6 (June 7, 2022): 1490–1500. https://doi.org/10.1093/ajcn/nqac055.

Baker, Laura D., Joann E. Manson, Stephen R. Rapp, Howard D. Sesso, Sarah A. Gaussoin, Sally A. Shumaker, and Mark A. Espeland. "Effects of Cocoa Extract and a Multivitamin on Cognitive Function: A Randomized Clinical Trial." *Alzheimer's & Dementia* (2022) 1–12. https://doi.org/10.1002/alz.12767.

The ecological fallacy is a surprisingly common problem in research. There are a number of interesting review articles about the concept in general and the classic example of Émile Durkheim's analysis of suicide rates in Catholics vs. Protestants.

Freedman, David A. "Ecological Inference and the Ecological Fallacy." *International Encyclopedia of the Social & Behavioral Sciences Technical Report* No. 549 (October 15, 1999). https://web.stanford.edu/class/ed260/freedman549.pdf.

Pai, Madhukar, and Jay S. Kaufman. "Case Studies of Bias in Real Life Epidemiologic Studies: Émile Durkheim and the Ecological Fallacy." McGill University. Accessed March 3, 2023. https://www

.teachepi.org/wp-content/uploads/OldTE/documents/courses/bfiles/The%20B%20Files_File3_Durkheim_Final_Complete.pdf.

Similarly, the association between height and health can be affected by the ecological fallacy if you look at national averages vs. the heights of individuals.

Samaras, T.T. "How Height Is Related to Our Health and Longevity: A Review." *Nutrition and Health* 21, no. 4 (October 2012): 247–61.

D'Avanzo, B., C. La Vecchia, and E. Negri. "Height and the Risk of Acute Myocardial Infarction in Italian Women." *Social Science & Medicine* (1982) 38, no. 1 (January 1994): 193–96. https://doi.org/10.1016/0277-9536(94)90315-8.

Silventoinen, Karri, Slobodan Zdravkovic, Axel Skytthe, Peter McCarron, Anne Maria Herskind, Markku Koskenvuo, Ulf de Faire, et al. "Association between Height and Coronary Heart Disease Mortality: A Prospective Study of 35,000 Twin Pairs." *American Journal of Epidemiology* 163, no. 7 (April 1, 2006): 615–21. https://doi.org/10.1093/aje/kwj081.

The graphic explaining the ecological fallacy is based on work from:

Weiss, Noel S., and Thomas D. Koepsell. *Epidemiologic Methods: Studying the Occurrence of Illness*. Second Edition. Oxford, New York: Oxford University Press, 2014.

CHAPTER 7: BREAKFAST'S THE MOST IMPORTANT MEAL OF THE DAY

The idea of breakfast being the most important meal of the day really does appear to have been the brainchild of Lenna Cooper, who coined the saying. The phrase was catchy enough to become a marketing slogan and changed how we perceive breakfast.

Hart, Karen. "The Reason People Believe Breakfast Is The Most Important Meal of the Day." *Mashed*, August 9, 2020. https://www.mashed.com/234731/the-reason-people-believe-breakfast-is-the-most-important-meal-of-the-day/.

Lennon, Troy. "How the Kellogg Brothers Influenced the Way Westerners Eat Breakfast." *Daily Telegraph*, August 7, 2018. https://www.dailytelegraph.com.au/news/how-the-kellogg-brothers-influenced-the-way-westerners-eat-breakfast/news-story/7478e97f93d98c466454ac970a88e9ad.

Kellogg had some rather unconventional ideas about sexual health, but whether he invented breakfast cereal to stop people from masturbating is somewhat dubious.

MacGuill, Dan. "Were Kellogg's Corn Flakes Created as an 'Anti-Masturbatory Morning Meal'? Snopes.Com." *Snopes*, August 16, 2019. https://www.snopes.com/fact-check/kelloggs-corn-flakes-masturbation/.

The introduction of bacon as the prototypical breakfast food was also a product of marketing. It was the handiwork of Edward Bernays, who was called the "Father of Public Relations" by the *New York Times*. While working for the Beech-Nut Packing Company, he developed a campaign to get Americans to eat a hardier breakfast of bacon and eggs. The results speak for themselves.

"Edward Bernays, 'Father of Public Relations' and Leader in Opinion Making, Dies at 103." *New York Times*, March 10, 1995. https://archive.nytimes.com/www.nytimes.com/books/98/08/16/specials/bernays-obit.html?_r=1.

Bernays, Edward L. "Interview with Edward L. Bernays." YouTube. Accessed May 4, 2023. https://www.youtube.com/watch?v=KLud EZpMjKU.

"Edward Bernays and Why We Eat Bacon for Breakfast." Braithwaite Communications, September 22, 2020. https://gobraithwaite.com /thinking/edward-bernays-and-why-we-eat-bacon-for-breakfast/.

Readers interested in the history and evolution of the meal we now call breakfast should consider reading:

Arndt Anderson, Heather. *Breakfast: A History*. AltaMira Studies in Food and Gastronomy. Lanham: AltaMira Press, a division of Rowman & Littlefield Publishers, Inc, 2013.

There has been quite a bit of research looking into the impact of breakfast programs in schools and whether or not they improve school performance. In general, they do, especially when you look at math scores. The benefit also tends to be more pronounced in children from low-income households, so it's a bit unclear whether breakfast itself is having a positive effect or whether better nutrition and food security just has an all-around improvement in children who are less well-off.

Imberman, Scott A., and Adriana D. Kugler. "The Effect of Providing Breakfast in Class on Student Performance." *Journal of Policy Analysis and Management* 33, no. 3 (2014): 669–99. https://doi.org /10.1002/pam.21759.

Adolphus, Katie, Clare L. Lawton, and Louise Dye. "The Effects of Breakfast on Behavior and Academic Performance in Children and Adolescents." *Frontiers in Human Neuroscience* 7 (August 8, 2013): 425. https://doi.org/10.3389/fnhum.2013.00425.

The Health Professionals Follow-up Study showed a link between men who reported skipping breakfast and subsequent heart disease. Other cohort studies showed similar findings.

Cahill, Leah E., Stephanie E. Chiuve, Rania A. Mekary, Majken K. Jensen, Alan J. Flint, Frank B. Hu, and Eric B. Rimm. "Prospective Study of Breakfast Eating and Incident Coronary Heart Disease in a Cohort of Male US Health Professionals." *Circulation* 128, no. 4 (July 23, 2013): 337–43. https://doi.org/10.1161/CIRCULATIONAHA.113.001474.

Ofori-Asenso, Richard, Alice J. Owen, and Danny Liew. "Skipping Breakfast and the Risk of Cardiovascular Disease and Death: A Systematic Review of Prospective Cohort Studies in Primary Prevention Settings." *Journal of Cardiovascular Development and Disease* 6, no. 3 (August 22, 2019): 30. https://doi.org/10.3390/jcdd6030030.

Some studies have shown that front-loading your meals, i.e., eating more earlier in the day than later at night, helps with weight loss. However, the data on this subject is inconsistent, and a 2018 review in the *British Medical Journal* found that there was no consistent benefit to eating breakfast as a means of weight loss. Also, a more recent study showed no increased weight loss when patients front-loaded their meals, though patients reported feeling less hungry.

Jakubowicz, Daniela, Maayan Barnea, Julio Wainstein, and Oren Froy. "High Caloric Intake at Breakfast vs. Dinner Differentially Influences Weight Loss of Overweight and Obese Women." *Obesity* 21, no. 12 (December 2013): 2504–12. https://doi.org/10.1002/oby.20460.

Sievert, Katherine, Sultana Monira Hussain, Matthew J. Page, Yuanyuan Wang, Harrison J. Hughes, Mary Malek, and Flavia

M. Cicuttini. "Effect of Breakfast on Weight and Energy Intake: Systematic Review and Meta-Analysis of Randomized Controlled Trials." *BMJ* 364 (January 30, 2019): l42. https://doi.org/10.1136/bmj.l42.

Ruddick-Collins, Leonie C., Peter J. Morgan, Claire L. Fyfe, Joao A. N. Filipe, Graham W. Horgan, Klaas R. Westerterp, Jonathan D. Johnston, and Alexandra M. Johnstone. "Timing of Daily Calorie Loading Affects Appetite and Hunger Responses without Changes in Energy Metabolism in Healthy Subjects with Obesity." *Cell Metabolism* 34, no. 10 (October 4, 2022): 1472-1485.e6. https://doi.org/10.1016/j.cmet.2022.08.001.

Studies have failed to show that intermittent fasting actually helps people lose weight. While earlier studies looked at alternate-day fasting and similar diet plans, more recent studies like the TREAT randomized trial and a more recent study in the *NEJM* showed that time-restricted eating didn't perform any better than a conventional diet. Some recent studies showing benefits in weight loss with intermittent fasting only had partial success, showing benefits in weight loss but not in reducing the percentage of body fat.

Jamshed, Humaira, Felicia L. Steger, David R. Bryan, Joshua S. Richman, Amy H. Warriner, Cody J. Hanick, Corby K. Martin, Sarah-Jeanne Salvy, and Courtney M. Peterson. "Effectiveness of Early Time-Restricted Eating for Weight Loss, Fat Loss, and Cardiometabolic Health in Adults with Obesity: A Randomized Clinical Trial." *JAMA Internal Medicine* 182, no. 9 (September 1, 2022): 953–62. https://doi.org/10.1001/jamainternmed.2022.3050.

Liu, Deying, Yan Huang, Chensihan Huang, Shunyu Yang, Xueyun Wei, Peizhen Zhang, Dan Guo, et al. "Calorie Restriction with or without Time-Restricted Eating in Weight Loss." *The New*

England Journal of Medicine 386, no. 16 (April 21, 2022): 1495–1504. https://doi.org/10.1056/NEJMoa2114833.

Lowe, Dylan A., Nancy Wu, Linnea Rohdin-Bibby, A. Holliston Moore, Nisa Kelly, Yong En Liu, Errol Philip, et al. "Effects of Time-Restricted Eating on Weight Loss and Other Metabolic Parameters in Women and Men with Overweight and Obesity: The TREAT Randomized Clinical Trial." *JAMA Internal Medicine* 180, no. 11 (November 1, 2020): 1491–99. https://doi.org/10.1001/jamainternmed.2020.4153.

Trepanowski, John F., Cynthia M. Kroeger, Adrienne Barnosky, Monica C. Klempel, Surabhi Bhutani, Kristin K. Hoddy, Kelsey Gabel, et al. "Effect of Alternate-Day Fasting on Weight Loss, Weight Maintenance, and Cardioprotection Among Metabolically Healthy Obese Adults: A Randomized Clinical Trial." *JAMA Internal Medicine* 177, no. 7 (July 1, 2017): 930–38. https://doi.org/10.1001/jamainternmed.2017.0936.

As for Jim's questions about study acronyms, the history of the evolution of study names is a fascinating read for anyone who wants to go on a bit of digression:

Labos, Christopher. "The Alphabet Soup of Clinical Trial Acronyms." *Medscape*, February 1, 2016. https://www.medscape.com/viewarticle/857543.

There are lots of books, blogs, articles, and editorials claiming we should eat more fat. They are too numerous to list here. A quick Google search yields many results and is a great way to start down a bottomless rabbit hole. The Twitter screenshots highlight how the PURE and ARIC studies generated mutually exclusive headlines on the same news site. Amazingly, no one seemed bothered by it.

The Telegraph. “Low-Carb, High-Fat Diets Could Knock Years Off Lifespan, 25-Year Study Suggests.” Twitter, August 17, 2018. https://twitter.com/Telegraph/status/1030343764085288960.

The Telegraph. “Low-Fat Diet Could Kill You, Major Study Shows.” Twitter, August 29, 2017. https://twitter.com/Telegraph/status/902512407875960833.

If you've never seen how life expectancy has changed in the past century, it is impressive to behold:

“United States: Life Expectancy 1860-2020.” Statista. Accessed March 4, 2023. https://www.statista.com/statistics/1040079/life-expectancy-united-states-all-time/.

The rise and fall of heart disease rates in the 20th century has many complex causes but was largely driven by smoking, dietary changes, and an overall change in lifestyle factors from the century before.

Dalen, James E., Joseph S. Alpert, Robert J. Goldberg, and Ronald S. Weinstein. “The Epidemic of the 20th Century: Coronary Heart Disease.” *The American Journal of Medicine* 127, no. 9 (September 2014): 807–12. https://doi.org/10.1016/j.amjmed.2014.04.015.

Anybody interested in an in-depth review of the history of cholesterol research can read the series of articles by Daniel Steinberg published in the *Journal of Lipid Research.*

Steinberg, Daniel. “Thematic Review Series: The Pathogenesis of Atherosclerosis. An Interpretive History of the Cholesterol Controversy: Part I.” *Journal of Lipid Research* 45, no. 9 (September 2004): 1583–93. https://doi.org/10.1194/jlr.R400003-JLR200.

Steinberg, Daniel. "Thematic Review Series: The Pathogenesis of Atherosclerosis. An Interpretive History of the Cholesterol Controversy: Part II: The Early Evidence Linking Hypercholesterolemia to Coronary Disease in Humans." *Journal of Lipid Research* 46, no. 2 (February 2005): 179–90. https://doi.org/10.1194/jlr.R400012-JLR200.

Steinberg, Daniel. "Thematic Review Series: The Pathogenesis of Atherosclerosis: An Interpretive History of the Cholesterol Controversy, Part III: Mechanistically Defining the Role of Hyperlipidemia." *Journal of Lipid Research* 46, no. 10 (October 2005): 2037–51. https://doi.org/10.1194/jlr.R500010-JLR200.

Steinberg, Daniel. "The Pathogenesis of Atherosclerosis. An Interpretive History of the Cholesterol Controversy, Part IV: The 1984 Coronary Primary Prevention Trial Ends It—Almost." *Journal of Lipid Research* 47, no. 1 (January 2006): 1–14. https://doi.org/10.1194/jlr.R500014-JLR200.

Steinberg, Daniel. "Thematic Review Series: The Pathogenesis of Atherosclerosis. An Interpretive History of the Cholesterol Controversy, Part V: The Discovery of the Statins and the End of the Controversy." *Journal of Lipid Research* 47, no. 7 (July 2006): 1339–51. https://doi.org/10.1194/jlr.R600009-JLR200.

Much has been written regarding the Seven Countries Study. For succinct reviews of what has been said and how much of it is true:

Blackburn, Henry. "Critiques of the Seven Countries Study and Ancel Keys." In *A History of Cardiovascular Disease Epidemiology*. University of Minnesota website. Accessed March 4, 2023. http://www.epi.umn.edu/cvdepi/multimedia/critiques-of-the-seven-countries-study-and-ancel-keys/.

Pett, Katherine, Joel Kahn, Walter Willet, and David Katz. "Ancel Keys and the Seven Countries Study: An Evidence-Based Response to Revisionist Histories WHITE PAPER." The True Health Initiative, 2017. https://www.truehealthinitiative.org/wp-content/uploads/2017/07/SCS-White-Paper.THI_.8-1-17.pdf.

The cover of *The Atlantic* from September 1989:

"Atlantic September 1989 Issue." *The Atlantic*. Accessed March 4, 2023. https://www.theatlantic.com/magazine/toc/1989/09/.

Despite many claims to the contrary, a review of studies involving statins and other cholesterol medications, which includes 49 trials with over 300,000 patients, shows that lowering the cholesterol in your blood reduces your cardiovascular risk. The more you lower cholesterol, the more your risk goes down.

Silverman, Michael G., Brian A. Ference, Kyungah Im, Stephen D. Wiviott, Robert P. Giugliano, Scott M. Grundy, Eugene Braunwald, and Marc S. Sabatine. "Association Between Lowering LDL-C and Cardiovascular Risk Reduction Among Different Therapeutic Interventions: A Systematic Review and Meta-Analysis." *JAMA* 316, no. 12 (September 27, 2016): 1289–97. https://doi.org/10.1001/jama.2016.13985.

Also, lowering your cholesterol to very low levels, a feat only possible recently with the newer cholesterol medications, has been shown to be safe with no signs of excess risk in this latest body of research.

Sabatine, Marc S., Stephen D. Wiviott, KyungAh Im, Sabina A. Murphy, and Robert P. Giugliano. "Efficacy and Safety of Further Lowering of Low-Density Lipoprotein Cholesterol in Patients Starting With Very Low Levels: A Meta-Analysis."

JAMA Cardiology 3, no. 9 (September 1, 2018): 823–28. https://doi.org/10.1001/jamacardio.2018.2258.

Current guidelines have moved away from recommending specific limits on how much cholesterol people eat in their diet. The current focus is on overall dietary patterns, which are easier to implement and have more evidence behind them. The current guidelines do a good job of summarizing the evidence and the evolution of guidelines with respect to cholesterol and egg consumption.

Carson J.A.S., Lichtenstein A.H., Anderson C.A.M., et al. "Dietary Cholesterol and Cardiovascular Risk: A Science Advisory from the American Heart Association." *Circulation* 141, no. 3 (January 21, 2020): e39-e53.

The difficulty in sorting through all the studies about eggs was laid out beautifully in a recent editorial that also has possibly one of the cleverest titles I have ever seen. It was written in response to the analysis linking eggs with cardiovascular disease.

Tobias, Deirdre K. "What Eggsactly Are We Asking Here? Unscrambling the Epidemiology of Eggs, Cholesterol, and Mortality." *Circulation* 145, no. 20 (May 17, 2022): 1521–23. https://doi.org/10.1161/CIRCULATIONAHA.122.059393.

Zhao, Bin, Demetrius Albanes, and Jiaqi Huang. "Response by Zhao et al. to Letter Regarding Article, 'Associations of Dietary Cholesterol, Serum Cholesterol, and Egg Consumption with Overall and Cause-Specific Mortality: Systematic Review and Updated Meta-Analysis.'" *Circulation* 146, no. 23 (December 6, 2022): e328–29. https://doi.org/10.1161/CIRCULATIONAHA.122.061978.

CHAPTER 8: CAFFEINE CAN TRIGGER HEART ATTACKS

The study suggested that there was an increased risk of heart attack in the first hour after drinking a cup of coffee.

Baylin, Ana, Sonia Hernandez-Diaz, Edmond K. Kabagambe, Xinia Siles, and Hannia Campos. "Transient Exposure to Coffee as a Trigger of a First Nonfatal Myocardial Infarction." *Epidemiology* (Cambridge, Mass.) 17, no. 5 (September 2006): 506–11. https://doi.org/10.1097/01.ede.0000229444.55718.96.

The case-crossover design used to study the heart attack risk after drinking a cup of coffee works best for situations where the risk factor you're interested in has an immediate and transient risk. In other words, risk goes up quickly when you start doing that thing and goes down just as quickly when you stop.

Maclure, M., and M. A. Mittleman. "Should We Use a Case-Crossover Design?" *Annual Review of Public Health* 21 (2000): 193–221. https://doi.org/10.1146/annurev.publhealth.21.1.193.

The accompanying editorial highlighted an important point about risk assessment. A relative risk can appear large while the absolute risk is quite small. This is why a 50 percent increase in risk can also be seen as needing two million cups of coffee to cause one heart attack.

Poole, Charles. "Coffee and Myocardial Infarction." *Epidemiology* (Cambridge, Mass.) 18, no. 4 (July 2007): 518–19. https://doi.org/10.1097/EDE.0b013e31806466e5.

You can do a similar calculation with the study showing that strawberries prevent heart attacks in women.

Cassidy, Aedín, Kenneth J. Mukamal, Lydia Liu, Mary Franz, A. Heather Eliassen, and Eric B. Rimm. "High Anthocyanin Intake Is Associated with a Reduced Risk of Myocardial Infarction in Young and Middle-Aged Women." *Circulation* 127, no. 2 (January 15, 2013): 188–96. https://doi.org/10.1161/CIRCULATIONAHA.112.122408.

In fact, most of the research around coffee suggests that it is either mildly beneficial or at the very least neutral when it comes to cardiovascular health.

Ding, Ming, Shilpa N. Bhupathiraju, Ambika Satija, Rob M. van Dam, and Frank B. Hu. "Long-Term Coffee Consumption and Risk of Cardiovascular Disease: A Systematic Review and a Dose-Response Meta-Analysis of Prospective Cohort Studies." *Circulation* 129, no. 6 (February 11, 2014): 643–59. https://doi.org/10.1161/CIRCULATIONAHA.113.005925.

However, the observed benefit might be due to reverse causation because people with pre-existing cardiac issues or risk factors, like high blood pressure, tend to avoid coffee and drink more decaffeinated coffee on average.

Hyppönen, Elina, and Ang Zhou. "Cardiovascular Symptoms Affect the Patterns of Habitual Coffee Consumption." *The American Journal of Clinical Nutrition* 114, no. 1 (July 1, 2021): 214–19. https://doi.org/10.1093/ajcn/nqab014.

That being said, excessive coffee consumption is associated with increased risk and drinking more than 10 cups of coffee per day does seems to increase your risk of sudden cardiac death.

Vreede-Swagemakers, J.J. de, A.P. Gorgels, M.P. Weijenberg, W.I. Dubois-Arbouw, B. Golombeck, J.W. van Ree, A. Knottnerus, and H.J. Wellens. "Risk Indicators for Out-of-Hospital Cardiac Arrest in Patients with Coronary Artery Disease." *Journal of Clinical Epidemiology* 52, no. 7 (July 1999): 601–07. https://doi.org/10.1016/s0895-4356(99)00044-x.

The FDA however recommends people limit themselves to a more modest four or five cups of coffee per day.

Commissioner, Office of the Commissioner. "Spilling the Beans: How Much Caffeine Is Too Much?" *Food and Drug Administration*, U.S. Food and Drug Administration website, February 12, 2021. https://www.fda.gov/consumers/consumer-updates/spilling-beans-how-much-caffeine-too-much.

Regular coffee consumption seems to be safe overall. The results of the CRAVE study, which were presented at the American Heart Association meeting in 2021, showed that drinking coffee didn't increase the risk of major arrhythmias, though possibly it did increase the risk of premature ventricular contractions which could cause symptoms of palpitations or skipped beats in people.

Wendling, Patrice. "CRAVE: No Spike in Atrial Arrhythmias Among Coffee Drinkers." *Medscape*, November 14, 2021. https://www.medscape.com/viewarticle/962909.

There have unfortunately been reports of caffeine overdoses and of young people dying suddenly from potential caffeine overdoses like in the case of Ohio teen Logan Stiner.

Taylor, Matthew. "Students 'Given Dose Equivalent to 300 Coffees' in Botched Test." *The Guardian*, January 25, 2017. https://www.theguardian.com/uk-news/2017/jan/25/students-caffeine-newcastle-crown-court-northumbria.

Lynch, Jamiel, and Debra Goldschmidt. "Teen Dies from Too Much Caffeine, Coroner Says." *CNN*, May 16, 2017. https://www.cnn.com/2017/05/15/health/teen-death-caffeine/index.html.

Anderson, Kristin. "Caffeine Overdose Ruled Cause of Prom King's Death." *USA TODAY*, July 1, 2014. https://www.usatoday.com/story/news/nation-now/2014/07/01/caffeine-overdose-prom-king/11863631/.

Sheets, Megan, and the Associated Press. "Amazon Is NOT Liable for Death of Ohio Teen Who Overdosed on Powdered Caffeine." *Daily Mail Online*, October 1, 2020. https://www.dailymail.co.uk/news/article-8795339/Court-Amazon-not-liable-teens-powdered-caffeine-death.html.

Which led to the FDA advisory and warning letters about concentrated caffeine powder:

Center for Food Safety and Applied Nutrition. "Guidance for Industry: Highly Concentrated Caffeine in Dietary Supplements." U.S. Food and Drug Administration website, September 21, 2022. https://www.fda.gov/regulatory-information/search-fda-guidance-documents/guidance-industry-highly-concentrated-caffeine-dietary-supplements.

CHAPTER 9: VITAMIN D IS THE CURE FOR EVERYTHING

Vitamin D deficiency is not as common as people think it is. If we take less than 30 nmol/L as deficiency, and between 30 and 50 nmol/L as vitamin D inadequacy, then the prevalence of these problems is around 5 percent and 18.3 percent in the U.S.

Herrick, Kirsten A., Renee J. Storandt, Joseph Afful, Christine M. Pfeiffer, Rosemary L. Schleicher, Jaime J. Gahche, and Nancy

Potischman. "Vitamin D Status in the United States, 2011–2014." *The American Journal of Clinical Nutrition* 110, no. 1 (July 1, 2019): 150–57. https://doi.org/10.1093/ajcn/nqz037.

Looker, A.C., Johnson C.L., Lacher D.A., et al. "Vitamin D Status: United States 2001–2006." NCHS Data Brief, no 59 (2011). Accessed March 3, 2023. https://www.cdc.gov/nchs/products/databriefs/db59.htm.

Did Tiny Tim have rickets or renal tubular acidosis? Believe it or not, the question has actually been discussed in the medical literature. Dickens never actually said what was wrong with him and since Tiny Tim is a creature of fiction and doesn't actually exist, the question can never be answered definitively. That hasn't stopped people from trying though.

Lewis, D.W. "What Was Wrong with Tiny Tim?" *American Journal of Diseases of Children* (1960) 146, no. 12 (December 1992): 1403–07. https://doi.org/10.1001/archpedi.1992.02160240013002.

Both Canada and the U.S. started fortifying food with vitamin D (and other nutrients) in the early part of the 20th century. The result was a correction of many of the nutritional deficiencies that occurred as a result of malnutrition. The laws vary slightly between the two countries as the U.S. has a voluntary supplementation framework.

Institute of Medicine (U.S.) Committee on Use of Dietary Reference Intakes in Nutrition Labeling. *Dietary Reference Intakes: Guiding Principles for Nutrition Labeling and Fortification*. Washington (DC): National Academies Press (U.S.), 2003. http://www.ncbi.nlm.nih.gov/books/NBK208881/.

Government of Canada, Public Works and Government Services Canada. "Canada Gazette, Part 2, Volume 156, Number 2: Marketing Authorization for Vitamin D in Milk, Goat's Milk and

Margarine." *Canada Gazette*, January 19, 2022. https://www.gazette.gc.ca/rp-pr/p2/2022/2022-01-19/html/sor-dors278-eng.html.

Center for Food Safety and Applied Nutrition. "Vitamin D for Milk and Milk Alternatives." U.S. Food and Drug Administration website, February 20, 2020. https://www.fda.gov/food/food-additives-petitions/vitamin-d-milk-and-milk-alternatives.

Most of what we perceive as aging is the fine wrinkles in our skin that come about as a result of sun exposure and photodamage. Sunscreen does more than just protect you from skin cancer. It also reduces the signs of aging and makes you look younger in the long term . . . at a much lower cost than many of the beauty products out there.

Hughes, Maria Celia B., Gail M. Williams, Peter Baker, and Adèle C. Green. "Sunscreen and Prevention of Skin Aging: A Randomized Trial." *Annals of Internal Medicine* 158, no. 11 (June 4, 2013): 781–90. https://doi.org/10.7326/0003-4819-158-11-201306040-00002.

It was the publication of the Women's Health Initiative that changed the narrative around hormone replacement therapy. Its findings showed that contrary to prior belief, routinely giving postmenopausal women hormones didn't in fact reduce cardiovascular events and possibly even increased harm.

Manson, JoAnn E., Rowan T. Chlebowski, Marcia L. Stefanick, Aaron K. Aragaki, Jacques E. Rossouw, Ross L. Prentice, Garnet Anderson, et al. "Menopausal Hormone Therapy and Health Outcomes during the Intervention and Extended Poststopping Phases of the Women's Health Initiative Randomized Trials." *JAMA* 310, no. 13 (October 2, 2013): 1353–68. https://doi.org/10.1001/jama.2013.278040.

Manson, JoAnn E., Judith Hsia, Karen C. Johnson, Jacques E. Rossouw, Annlouise R. Assaf, Norman L. Lasser, Maurizio Trevisan, et al.

"Estrogen Plus Progestin and the Risk of Coronary Heart Disease." *The New England Journal of Medicine* 349, no. 6 (August 7, 2003): 523–34. https://doi.org/10.1056/NEJMoa030808.

There was a time when everybody was taking vitamin D to prevent everything from heart disease to cancer. But as one editorial put it, "Then came the randomized trials." Even a study testing vitamin D supplementation in vitamin D–deficient children didn't show a positive result.

Lucas, Anika, and Myles Wolf. "Vitamin D and Health Outcomes: Then Came the Randomized Clinical Trials." *JAMA* 322, no. 19 (November 19, 2019): 1866–68. https://doi.org/10.1001/jama.2019.17302.

Manson, JoAnn E., Nancy R. Cook, I-Min Lee, William Christen, Shari S. Bassuk, Samia Mora, Heike Gibson, et al. "Vitamin D Supplements and Prevention of Cancer and Cardiovascular Disease." *New England Journal of Medicine* 380, no. 1 (January 3, 2019): 33–44. https://doi.org/10.1056/NEJMoa1809944.

Urashima, Mitsuyoshi, Hironori Ohdaira, Taisuke Akutsu, Shinya Okada, Masashi Yoshida, Masaki Kitajima, and Yutaka Suzuki. "Effect of Vitamin D Supplementation on Relapse-Free Survival among Patients with Digestive Tract Cancers: The AMATERASU Randomized Clinical Trial." *JAMA* 321, no. 14 (April 9, 2019): 1361–69. https://doi.org/10.1001/jama.2019.2210.

Pittas, Anastassios G., Bess Dawson-Hughes, Patricia Sheehan, James H. Ware, William C. Knowler, Vanita R. Aroda, Irwin Brodsky, et al. "Vitamin D Supplementation and Prevention of Type 2 Diabetes." *The New England Journal of Medicine* 381, no. 6 (August 8, 2019): 520–30. https://doi.org/10.1056/NEJMoa1900906.

Brustad, Nicklas, Anders U. Eliasen, Jakob Stokholm, Klaus Bønnelykke, Hans Bisgaard, and Bo L. Chawes. "High-Dose Vitamin D Supplementation During Pregnancy and Asthma in Offspring at the Age of 6 Years." *JAMA* 321, no. 10 (March 12, 2019): 1003–05. https://doi.org/10.1001/jama.2019.0052.

De Boer, Ian H., Leila R. Zelnick, John Ruzinski, Georgina Friedenberg, Julie Duszlak, Vadim Y. Bubes, Andrew N. Hoofnagle, et al. "Effect of Vitamin D and Omega-3 Fatty Acid Supplementation on Kidney Function in Patients With Type 2 Diabetes: A Randomized Clinical Trial." *JAMA* 322, no. 19 (November 19, 2019): 1899–1909. https://doi.org/10.1001/jama.2019.17380.

Barbarawi, Mahmoud, Babikir Kheiri, Yazan Zayed, Owais Barbarawi, Harsukh Dhillon, Bakr Swaid, Anitha Yelangi, et al. "Vitamin D Supplementation and Cardiovascular Disease Risks in More Than 83,000 Individuals in 21 Randomized Clinical Trials: A Meta-Analysis." *JAMA Cardiology* 4, no. 8 (August 1, 2019): 765–76. https://doi.org/10.1001/jamacardio.2019.1870.

Forno, Erick, Leonard B. Bacharier, Wanda Phipatanakul, Theresa W. Guilbert, Michael D. Cabana, Kristie Ross, Ronina Covar, et al. "Effect of Vitamin D3 Supplementation on Severe Asthma Exacerbations in Children with Asthma and Low Vitamin D Levels: The VDKA Randomized Clinical Trial." *JAMA* 324, no. 8 (August 25, 2020): 752–60. https://doi.org/10.1001/jama.2020.12384.

Bischoff-Ferrari, Heike A., Bruno Vellas, René Rizzoli, Reto W. Kressig, José A. P. da Silva, Michael Blauth, David T. Felson, et al. "Effect of Vitamin D Supplementation, Omega-3 Fatty Acid Supplementation, or a Strength-Training Exercise Program on Clinical Outcomes in Older Adults: The DO-HEALTH

Randomized Clinical Trial." *JAMA* 324, no. 18 (November 10, 2020): 1855–68. https://doi.org/10.1001/jama.2020.16909.

Okereke, Olivia I., Charles F. Reynolds, David Mischoulon, Grace Chang, Chirag M. Vyas, Nancy R. Cook, Alison Weinberg, et al. "Effect of Long-Term Vitamin D3 Supplementation vs. Placebo on Risk of Depression or Clinically Relevant Depressive Symptoms and on Change in Mood Scores: A Randomized Clinical Trial." *JAMA* 324, no. 5 (August 4, 2020): 471–80. https://doi.org/10.1001/jama.2020.10224.

Burt, Lauren A., Emma O. Billington, Marianne S. Rose, Duncan A. Raymond, David A. Hanley, and Steven K. Boyd. "Effect of High-Dose Vitamin D Supplementation on Volumetric Bone Density and Bone Strength: A Randomized Clinical Trial." *JAMA* 322, no. 8 (August 27, 2019): 736–45. https://doi.org/10.1001/jama.2019.11889.

Ganmaa, Davaasambuu, Sabri Bromage, Polyna Khudyakov, Sumiya Erdenenbaatar, Baigal Delgererekh, and Adrian R. Martineau. "Influence of Vitamin D Supplementation on Growth, Body Composition, and Pubertal Development Among School-Aged Children in an Area With a High Prevalence of Vitamin D Deficiency: A Randomized Clinical Trial." *JAMA Pediatrics* 177, no. 1 (January 1, 2023): 32–41. https://doi.org/10.1001/jamapediatrics.2022.4581.

Medscape's F. Perry Wilson had a great video capsule summarizing the latest vitamin D research and explaining why it's probably a marker of disease rather than the cause.

Wilson, F. Perry. "The Surprising Failure of Vitamin D in Deficient Kids." *Medscape*, November 29, 2022. https://www.medscape.com/viewarticle/984626.

This book is also available as a Global Certified Accessible™ (GCA) ebook. ECW Press's ebooks are screen reader friendly and are built to meet the needs of those who are unable to read standard print due to blindness, low vision, dyslexia, or a physical disability.

At ECW Press, we want you to enjoy our books in whatever format you like. If you've bought a print copy just send an email to ebook@ecwpress.com and include:

- the book title
- the name of the store where you purchased it
- a screenshot or picture of your order/receipt number and your name
- your preference of file type: PDF (for desktop reading), ePub (for a phone/tablet, Kobo, or Nook), mobi (for Kindle)

A real person will respond to your email with your ebook attached. Please note this offer is only for copies bought for personal use and does not apply to school or library copies.

Thank you for supporting an independently owned Canadian publisher with your purchase!